The Automobile Revolution

Danielle Attias
Editor

The Automobile Revolution

Towards a New Electro-Mobility Paradigm

Springer

Editor
Danielle Attias
Laboratoire Genie Industriel,
CentraleSupélec
Université Paris-Saclay
Chatenay-Malabry
France

ISBN 978-3-319-45837-3 ISBN 978-3-319-45838-0 (eBook)
DOI 10.1007/978-3-319-45838-0

Library of Congress Control Number: 2016952508

Printed on acid-free paper

This Springer imprint is published by Springer Nature
The registered company is Springer International Publishing AG
The registered company address is: Gewerbestrasse 11, 6330 Cham, Switzerland

Preface

Electro mobility is a general global trend, since it corresponds to converging expectations regarding the environment and society at the same time as institutions, communities and companies undergo a digital transformation. Electric vehicles will be connected vehicles, once again due to the increasingly clear convergence between information and energy networks, opening the way towards shared objectives of autonomy, reliability and security.

It is already obvious that the new generation of electric vehicles cannot be devised, designed and manufactured in the same way as conventional electrified vehicles, not least because the current revolution in terms of how we perceive vehicles and their new uses requires a totally new approach to electric vehicles' architecture, structure, modularity and operation and their interactions with the outside world, until they gradually become autonomous.

At the same time, carmakers are facing rapid technological change, especially in terms of equipment, energy storage, broadband communications and artificial intelligence, sometimes within time frames that are much shorter than the duration of a project. As a result, they have to design upgradeable electric cars capable of easily integrating new physical and software technologies. Innovation therefore becomes crucial to compete with other carmakers, as well as with "new entrants" from the domain of information technology.

It goes without saying that for traditional carmakers, which have nevertheless experienced one or several industrial revolutions including the arrival of robots in their factories, the current revolution is totally different. This is partly because it calls into question the actual purpose of their production as well as their positioning in mobility value chains, where services are likely to increasingly represent a larger share than products.

New economic models for electro mobility are emerging; major changes are taking place in the car industry, with a rising proportion of mobility service activities. What will the car of the future be like? Hybrid, intelligent, autonomous, driverless, but without doubt a shared car for user communities whose primary requirements are time, security and travel comfort. Can the autonomous car meet all of these challenges? Clearly, carmakers will need to outsource information and communication activities to increase their companies' in-house flexibility and agility. This is because it is vital to create the conditions for fostering collective intelligence and to form a learning/unlearning hotbed to stimulate innovation. For traditional carmakers, the future will involve a mobility revolution combined with a cultural revolution of the entire automotive industry. However, carmakers should retain their own, clearly defined identity as they set out to tackle these new challenges.

In my view, the work presented in this publication makes a pertinent contribution to the debate and thinking process behind the current revolution.

Paris, France Sylvain Allano

Acknowledgments

This book is the result of research carried out over several years within the Armand Peugeot Research Chair. I would like to thank Mr. Sylvain Allano, Chief Scientist for Future Technologies at PSA Peugeot-Citroën, for his expertise, his confidence and his precious help to continue our work, to Mr. Jean-Marc Mousset, Head of Industrial Academy and Training Projects, for his enlightening points of view and his requirement from us 'to outperform', and also to all those teams of PSA Peugeot-Citroën for the time they spent with us, helping us through fruitful exchanges and enriching our knowledge.

My gratitude also goes to Hakim Idjis, Researcher at the Industrial Engineering Laboratory at CentraleSupélec, who helped by his invaluable assistance in this writing project.

Contents

Chapter 1
Introduction

Danielle Attias

Of one thing we can be certain: the evolution towards electro mobility will be a revolution. Imagining the mobility needs of tomorrow means taking into account multiple factors regarding energy, the environment, socio-economics, politics and technology, in a context of social transformation that stretches far beyond the automotive industry.

There are often contradictory hypotheses that attempt to explain the evolution of mobility over the next 20 years. Technological and economic uncertainties aside, the most widely accepted claim is that the development of electro mobility will not only cause a complete disruption of urban mobility services for the movement of users, but also a change in behavior and the positions of the different players (manufacturers, energy companies, computer scientists and public authorities).

Far from wanting to pretend to answer all these questions which we do not yet measure the extent, we have chosen to lead a multidisciplinary reflection that allows us to highlight different issues related to various disciplines. Economics, management science, social sciences and engineering sciences are summoned in this book in order to enlighten this new concept of electro mobility. This book is organized in three main parts.

The first part of this book questions the emergence of a new mobility paradigm. What are the signs which highlight this disruption? How do we analyze the upheaval of the automotive industry and why does the autonomous car appear now? What will be the place of each actor and newcomers in the transformation of the value chain? Why does the government invest worldwide this new field of mobility and what are the economic, social and financial implications of their actions?

D. Attias (✉)
Industrial Engineering Research Department/Laboratoire Génie Industriel, CentraleSupélec, Université Paris-Saclay, Grande Voie des Vignes, 92290 Chatenay-Malabry, France
e-mail: danielle.attias@ecp.fr

D. Attias (ed.), *The Automobile Revolution*, DOI 10.1007/978-3-319-45838-0_1

To illuminate the present, analyzing the different stages of the automotive industry is necessary because, in the words of the French philosopher Jean-Paul Sartre, "it is the present that tells us if the past is still alive or dead". While the industry has experienced some inertia in the past, we have witnessed an acceleration for the last 15 years, with the emergence of new types of vehicles, new services, new needs and uses that break with the traditional car model. In this context, carmakers live with greater complexity and uncertainty in the orientation of their strategic choices. The second chapter describes all these changes—from automotive to self-mobility—and covers all aspects of topics investigated in depth in the following ones.

The transformation of the value chain and value proposition within the new mobility paradigm is tackled in the third chapter. It aims to examine, with an integrated interdisciplinary approach, why current ecological, economic and social changes will affect individual mobility, and if innovation can provide an answer to these challenges. It also compiles current approaches and business models of OEMs together with upcoming players in this transformed value chain.

The fourth chapter investigates the impact and involvement of urban policies in the mobility ecosystem. The reflection focuses on the role played by the public actor in the transformation of the automotive industry to the ecosystem of electro mobility. To illustrate this new urban paradigm, examples of innovative urban policies in megacities in Asia, Europe and the USA will be presented and analyzed in the context of ongoing changes.

The second part of this book addresses the links between the actors in the automotive industry and the users from a systemic and prospective viewpoint. The question of uses and new vehicles in an urban eco system under reconstruction will also be taken into consideration. Concepts such as smart cities and autonomous vehicles are introduced and studied in depth. The question is the one of this automobile revolution truly leading us towards a societal revolution.

First, the study of alliances and new partnerships in the automotive industry shows that the traditional economic model of this sector is down and new business models must accompany this change of paradigm. The purpose of chapter five to analyze how carmakers are forced to build new partnerships to *survive*, how should they anticipate changes and make wise choices. Because the future mobility ecosystem that is emerging is complex and requires *ambidextrous* positioning such as the one of coopetition developed here.

Another dimension that must be taken into account in this automotive revolution, is urban agglomerations or mega-cities in which the cars of tomorrow will develop. Chapter six explores scenarios of future *smart cities* that will have to respond to many issues (energy, mobility, housing, health, labor and infrastructure). Future mobility requires rethinking the urban space in terms of infrastructure and data management to optimize transport organization. The new ecosystem described in this chapter shows the emergence of these smart cities, new urban models facing major disruptive challenges. From the ongoing technological innovations in the field of mobility (the autonomous car) to the inhabitants lifestyle whose relationship on movement and work is changing radically, these smart cities are also an incredible challenge for the traditional automotive world and compels them to develop new business models.

Chapter seven takes the paradigm shift one step further, by investigating the autonomous car. It considers the self-driving car as the final step of an underlying trend in vehicle automation. But the autonomous car creates a radical break with other mobility modes and lifestyles, because it can be thought as a *shared* car. Imagine the residents of a neighborhood jointly using this *mobile object on four wheels*, sharing it and inventing new uses together and thus forming a new social link. However, many obstacles remain to be overcome—social, legal and financial issues to name a few, and need to be taken into account by carmakers. The latter are facing the arrival of new players like Google and Uber, players who are ready to create disruptive business models.

After describing the environmental, socio-economic and political context of this new mobility paradigm, the third and last part of this book is the technical-economic one. It tackles the technological barriers to overcome in this transition, but also the potential opportunities to seize. The transitional period where there will be coexisting thermal, hybrid, electric and then autonomous cars, raises many issues to carmakers such as plugging the electric vehicles (EV) to the existing energy infrastructure, the compliance with regulations in terms of recycling and lowering the EV cost so it can be competitive with thermal ones. Our focus is primarily technological and economic as these two points are proving essential to achieve this transition.

The eighth chapter studies the key component accompanying vehicle electrification, namely the lithium-ion battery. The battery makes up to 30 % of the EV cost and today, EVs are expensive compared to thermal ones. Furthermore, EV batteries are subject to recycling regulation in Europe, imposing on carmakers the recycling of at least half the battery weight which might be costly for batteries that usually weight more than 200 kg. So, the stake for carmakers is to make sure that EVs are still competitive for costumers while complying with the different regulations. Given this context, the battery end-of-life recovery impact is assessed in order to determine whether it will be a cost or a profit source for automakers.

With the same objective of managing the technological barriers in this transitional period, and also the opportunity to reduce the EV cost, the ninth and last chapter investigates the opportunity to use electric vehicles for grid services. Today, there is no clear framework on how EVs can participate in such services. In addition to this, the diversity of organizational forms or governance structures in the Transmission System Operators (TSO) makes it even more complicated. Based on a detailed analysis of the rules implemented by some representative TSOs, the authors identified the best aggregation rules and payment schemes for grid services, which allows the proposition of an ideal TSO organization. Policy recommendations are also provided.

To rethink future mobility is to rethink future cities and the future lifestyles of its inhabitants. The story of electro mobility is being written day by day.

Part I
The Emergence of a New Mobility Paradigm

Chapter 2
The Automobile World in a State of Change

From the Automobile to the Concept of Auto-Mobility

Danielle Attias

Abstract The automotive industry is undergoing a profound transformation with the arrival of new types of vehicles, services, requirements and uses that break away from the traditional model of cars. The very nature of mobility is changing. User behavior is shifting; social symbols connected to vehicle prestige and journey priorities are not the same. The aspiration attached to private cars is also in the process of changing. A look back at the history of the automobile reveals how our relationship to the car object is altering. It also reveals the need for a different approach to the current concept of mobility. New services (car-sharing, carpooling, self-service cars) are making a significant contribution to the emergence of a new ecosystem. The resulting industrial, economic and ecological situation involves all of the traditional stakeholders in the sector, i.e. car manufacturers, parts manufacturers, recyclers, energy and fuel suppliers, as well as market newcomers like engineering, computing and communications companies. In this new mobility ecosystem, manufacturers are changing their strategies. They find themselves obliged to reinvent an entire industrial model and work with companies far removed from their core business. Are these strategies sufficient to respond to the economic transformations affecting the automotive industry? Is this mobility transformation the basis of a paradigm shift? And lastly, how can public authorities accompany this technological and societal rupture and establish mobility that is ecological, responsible and more shared?

Keywords New needs · Uses and services of mobility · New mobility ecosystem and societal rupture · Paradigm shift for the automotive industry

D. Attias (✉)
Industrial Engineering Research Department/Laboratoire Génie Industriel, CentraleSupélec, Université Paris-Saclay, Grande Voie des Vignes, 92290 Chatenay-Malabry, France
e-mail: danielle.attias@ecp.fr

D. Attias (ed.), *The Automobile Revolution*, DOI 10.1007/978-3-319-45838-0_2

2.1 Introduction

To better understand the current change in mobility, we need to take a close look at the major historical periods of the automotive industry—from Fordism to the manufacture of driverless cars—and at how vehicles were devised at these different stages. Mobility is linked to social organization and can profoundly modify the attitudes of modern society to time, space and inter-personal relationships. Thus, by taking a historical perspective, we show that the very *nature* of mobility is in the process of changing.

The socio-economic realities of the last decade has accelerated a change in the car's image—with a focus on downsides like noise, pollution, excessive purchase and usage costs, time wasted on journeys, etc. A car becomes an *auto-immobile* when subject to urban congestion.

New behavior patterns, uses and mobility requirements are emerging that no longer view vehicles as an object of pleasure, freedom and social mobility, and in which usage can substitute ownership. Users today are not only seeking more rational management of journey times and mobility costs, but ultimately, they want to choose mobility rather than be subjected to it.

The first issue concerns the characterization of this change. Does it involve an evolution in the mobility system, a change to the technological business model, or the emergence of a new paradigm? These questions are decisive because they determine the directions of the different scenarios of technico-economic or societal change.

The first part will outline the development stages of the automobile which, according to (Oliver Wyman 2015), prefigure "*manufacturers' inevitable preparation for radical change*" (Sect. 2.2). The second part shows that technological rupture is accompanied by a change in usage and behavior involving new positioning for all mobility stakeholders (Sect. 2.3).

2.2 Emergence of a New Mobility Paradigm?

2.2.1 A Context of Rupture

In a world inhabited by close to 7 billion people, 800 million vehicles exist. "*Global production has returned to pre-crisis levels, reaching 90.6 million vehicles in 2015*" (Freyssenet 2015), with a steadily rising global trend of around 3 % per year (Observatoire Cetelem 2015). However, we only spend about 15 days in our cars each year, and a whole year of our lives looking for a parking place (Klappenecker et al. 2014). Car pollution has reached record levels in major urban areas and, if the European Commission applies a tax of €90 per gram of CO_2 in 2020, it will be twice as expensive as a gram of gold at current bullion prices. Although governments and policies generally strongly support the automotive industry, profitability is only about 3–4 % compared to double-digits in many other sectors.

Lastly, purchasers of cars in richer countries no longer view them as objects of social standing to own and display, but as a mobility item to be shared (Schaefers 2013). Thus, the authors of the final report on innovative car uses and new mobility (CGDD et al. 2016) point out that, "*Car demand is primarily a demand for mobility: if it can be satisfied without a car or with fewer kilometers, households do not hesitate to use other services that bring down their costs.*"

This context, which involves paradoxes and hesitations at every decision-making level (macro-meso-micro), results in a questioning of the automotive product, a review of mobility paradigms, and a reflection on the future of auto-mobility. The golden age of car manufacturing in the Triad countries is over. The new mass markets are in Asia and emerging countries, and forecasts for 2020 confirm this trend in several studies (OECD 2013; KPMG 2015), backed up by a PWC Autofacts report (PWC 2014): "*Car sale forecasts show that most growth will take place in emerging markets. The growth rate forecast up to 2020 is on average 3 % per year for OECD countries and 9 % for emerging countries... of which China and India are likely to represent 40 % of vehicle demand.*".

Nevertheless, buyers in emerging countries hesitate between owning "the same cars as rich people" or obtaining a means of transport that perfectly corresponds to their local realities (Qian and Soopramanien 2015).

The car is a century-old, mainstream product with significant industrial and technological content, but whose development can only be comprehended from a historical, geopolitical and social perspective. To determine the vehicle of the future and which industry will produce it, the question is whether it will involve continued incremental innovation or a conceptual shift (Boyer and Freyssenet 2006). For this reason, we have chosen to set out the development stages of the automobile to identify the levels of innovation relating to the car's different functions.

2.2.2 The Car, a Means of Transport to Be Seen

In its first stage, the motor car was a European-American concept defined as a personal or family tool used for traveling from one point to another, as best as possible, as fast as possible, and with optimum safety, comfort, ergonomics, etc. The first level of innovation therefore involves optimizing the car as an object of personal or family transportation. For a century, the car's technical fundamentals (i.e. 4 wheels, a motor and a body) changed little. Cars were massively produced in the Triad countries, and constituted a basic component of social achievement through personal mobility. The automobile was an object of social recognition and value to be owned. This positive image was accompanied by an idea of pleasure and personal freedom. The automobile was an object "to be seen" (cf. the Concours d'Elégance) with incremental innovations mostly centered on design and comfort.

This first stage has been taking place since the 1990s in emerging countries in Asia, with a range of innovations aimed at making the car a symbol of social achievement as well as an efficient means of travel (Artus et al. 2011).

However, denser urbanization and the development of megacities are increasing the vehicle stock and creating serious ecological and economic issues (urban pollution, dangerous driving, traffic jams, etc.). These dysfunctions have for a long time been the subject of studies by public institutions (in France mainly by ADEME—French Environment and Energy Management Agency (ADEME 2012); in the USA by the EPA), with analysis from a forecasting angle. Current R&D also takes these dysfunctions into account, perceiving tomorrow's vehicle in terms of energy savings and greenhouse gas emissions reduction (Plassat 2012).

Whether internal combustion, electric or hybrid, cars as a means of transport are still conceived with the same idea of mobility. The modes of innovation are incremental; they center on ergonomics to suit lifestyles and on reducing general energy costs, CO_2 emissions and total cost of ownership. Car manufacturers are the only ones to decide on these modes.

2.2.3 *The Car, a Component of Mobility 2.0*

The second stage considers the automobile as an element of mobility with which users "do more than move"[1] No longer simply an object of transport, the car takes on a concept of interactive mobility that creates a specific relationship between the driver and space and an information and communication network. Over and above transportation, the priority for drivers is no longer to be seen in their car, but to remain constantly connected to their information networks (GPS, WiFi, 3G etc.).

In this situation, product innovation is no longer simply incremental. It includes services that close the gap between individual transportation and collective transportation and encourage shared usage like rental, car sharing and carpooling. As a reminder, car sharing increases the productivity of a vehicle by maximizing its usage time, whereas carpooling maximizes its occupancy.

The immediate consequence is that the mobility range is reconfigured and diversified to become multimodal (people move from car to train, from train to bicycle or another (hire) car, while their own car might be used for a journey by another user who might also share it with others, etc.). We are gradually moving towards personal public transport (PPT). We are moving away from a set-up in which accessing mobility, i.e. freedom, involves owning a vehicle, and towards transportation in which journey satisfaction is determined by diverse means, easy access and low cost. And also obviously moving from a business model based on selling a product, towards a model based on selling a set of services whose value depends on the usage time of a vehicle, the volume used in a vehicle, or the distance travelled.

[1]The expression "mobility 2.0" is a reference to Web 2.0, which "allows users to do more than retrieve information". The expression was coined by Dale Dougherty in 2004 when he suggested that the Web was in a phase of renaissance or mutation, with a paradigm shift and evolving models.

Mobility 2.0 raises new strategic questions for today's car manufacturers. Firstly, it has an impact on their production volumes, because a shared car replaces 4–8 cars (IEA 2012). Next, the resources and skills indispensable for developing this mobility are not at all restricted to the traditional automotive industry. They depend on complementary operators (e.g. providers of complementary transport services, urban planners and infrastructure designers) and especially on the development of ICTs, which are essential for interconnections between transport means and users. More than ever before, manufacturers' innovation efforts are therefore linked to those of these operators and to changes in mobility patterns.

The traditional motorist no longer exists. Today's drivers are users who are part of a community; they are service users. They do not choose their vehicle based on material features or immaterial characteristics (status, size, brand, price, etc.); they are looking for a means of mobility within a community of practice. This is illustrated by the success of specialist carpooling companies like Blablacar in France and Lift and Uber in the USA. In less than 10 years, the number of Blablacar users has gone from a few thousand to 3 million people. The company employs around thirty staff and has subsidiaries in Europe (Spain, Portugal, Germany, Italy, etc.). The same trend can be seen in the USA, where the number of users tripled (450,000–1,400,000) from 2010 to 2014. The value proposition of these carpooling companies' centers on lower costs, reduced CO_2 emissions ["*10 million journeys and 700 thousand tons of* CO_2 *saved*" (BlaBlaCar 2016)], in addition to a more sociable trip.

Overall, mobility 2.0 means radical upcoming changes in the way that means of transport are used. This usage is connected to new social practices involving communities and social networks (Facebook, LinkedIn), the need for information and freedom (unrestricted internet), and indisputably free, shared, immediate facilities (open source and zapping). In this situation, manufacturers of "personal cars" produce goods that are no longer a priority for mobility 2.0 users. What is more, innovations to this product now depend on other contributors to mobility 2.0 goods and services.

2.2.4 The Vehicle of the Future: Intelligent Transportation in a "de-Mobility" Context

The third stage of the evolution of automobile innovation has yet to take place. It envisions mobility in general and the car as an intelligent component of our mobile and social lives (Nazem et al. 2011). The idea is that the mobility-car tandem is outdated in as much as we already live in a society in which life is so mobile that it compels *de-mobility* (i.e. not making journeys): work becomes tele-work; teaching takes the form of distance learning; meetings and discussions take place on social networks; culture and leisure are transformed or reinterpreted by mobility. Individual relationships with space and territory are changing the specific relationship between motorists and space and communication (as in mobility 2.0) are added their relationships with time and quality of life. This involves conceiving a particular way of life in which journeys are free from constraints and transport time creates value. How will today's automobile manufacturers fit into these new innovation modes?

2.3 From Automobile to Auto-Mobility

2.3.1 The Automotive Industry Is Reinventing Itself

According to Freyssenet (2009), the automotive industry needs to reflect deeply on how it can convert to meet technological and energy challenges and deal with the major changes in usage and types of mobility. Several questions arise regarding manufacturers' capacity to first tackle these new environmental challenges. According to a study produced by KPMG on the automotive industry in 2015 (KPMG 2015), the automotive value chain is undergoing considerable change and the battle to control it is only just beginning. The automotive industry is organized around a structured, homogenous supply chain that evolves in line with manufacturers' externalizing policies. Its sequence organization is firmly anchored in the hierarchies between suppliers, manufacturers and distributors (Frigant and Jullien 2014). Until now, the efficiency of this supply chain has resulted from its simplicity and stability.

Each stakeholder's role is the direct outcome of strategic decisions made by the manufacturers who control the main resources. Power games are therefore likely and, even though the most dependent stakeholders are critical of the value-sharing rules, they change only very little on the margins. However, as part of our investigation into how the automotive industry's technico-economic paradigm is changing, like Jullien and Pardi (2013) we wonder, "*Who will control the value of the future automobile products*?".

The three stages of the automobile's development that we have described involve several levels of innovation. Some are incremental, others are radical. Although there is some overlapping between levels, does the thesis hold of a technico-economic paradigm shift in a context of rupture in the automotive industry (Freyssenet 2011)? For Womack et al. (1991), who are researchers at MIT and authors of a seminal work, *The Machine that Changed the World*, the automotive industry will be saved by *technology*. Yet, the economic and global space is undergoing a fundamental reorganization and, as pointed out by Boyer and Freyssenet (2006), "*Countries are emerging. The relevant profit strategies for this new situation are not yet clearly defined, nor are the productive organizations.*" They go further by stating that, "*A century of automotive techniques suggests that no manufacturer has succeeded in totally revolutionizing the product through technology alone, apart from the ground-breaking Ford T episode…*".

The significance of this rupture should not just be judged from a historical perspective, but everything indicates that mobility as a whole is likely to be reinvented (Attias 2013). We have invented "cars for living" (Renault's "voitures à vivre"), and we accept the idea that tomorrow's mobility will transform relationships with others, lifestyles and the sources of value creation.

2.3.2 What Production Model for Auto Mobility 2.0

Manufacturers are obliged to evaluate their own skills and build new cost-reduction and commercialization strategies adapted to customers' purchase decisions. Since technology is increasingly complex and has shorter life cycles, it seems unlikely that a car manufacturer could on its own have sufficient financial resources and expertise to take the lead. In fact, alliances and joint developments already exist with partners from other industries (e.g. suppliers of batteries and electrical parts, technological information services for connected vehicles). Examples of how the global automotive industry is being reorganized and the need to build alliances further afield include cooperation between the French car manufacturer PSA-Peugeot-Citroën and the American computer company IBM; the partnership between the Spanish manufacturer Seat and the Korean telecommunications company Samsung; the alliance between the suppliers TomTom and Bosch; and the Chinese car manufacturer SAIC and the e-commerce company Alibaba.

Thus, numerous stakeholders are emerging that can be mobilized to work on the same project, i.e. the electromobility economy. In addition to the usual, directly concerned stakeholders, newcomers include rental companies, public transport operators, and energy providers that, alongside IT companies, are the emerging players in the game. They participate in drawing up these new business models and in other forms of cooperation that have dominated up till now.

By studying different strategies, some authors (Arjaliès and Ponssard 2010) have shown that, in situations similar to that of the automotive industry, two approaches are implemented. The so-called "conformity" approach is different from the "opportunity" approach. In the opportunity approach, a company's proactive attitude can give it a competitive advantage.

This is the case for some manufacturers that radically change direction, technological program and, ultimately, their image. Carlos Tavares, CEO of PSA Peugeot-Citroën, made a speech in January 2016 that broke with the traditional image of a car manufacturer and announced a major strategic shift, with the launch of electric/autonomous cars when the company's core business is internal combustion/hybrid cars. In assuming this approach, a company that knows how to take into account structural factors inherent to its business sector, such supply network, connection with suppliers, and the necessary cooperations and alliances, can hope to obtain a leading position in its area of activity (Arjaliès et al. 2011). The Chap. 4, on car manufacturer strategies, analyzes this major change and the economic, financial and social implications for all stakeholders.

If the auto-mobility 2.0 industry is to survive, should it place innovation at the heart of its system? The analysis made by the Austrian economist Schumpeter in his writings on innovation sheds light on the relationship between systems and innovation (Deblock 2012). For the author, new structures or organizations need to be created to truly permit disruptive innovation. Schumpeter shows that when innovative companies become more technocratic, they are no longer capable of challenging existing paradigms, knowledge and production modes. They then lose their capacity to come up with innovative products. For Schumpeter, "*Innovation is multi-dimensional and also concerns changes in the company's in-house organization and its ecosystem.*"

The innovation issue obviously depends on corporate strategy, but it highlights the difficult choices facing car manufacturers. Back in 2006, Boyer and Freyssenet questioned the "right" strategy for manufacturers to adopt. "*Should they anticipate some radical innovations in terms of technology (e.g. combining mechanics, electronics and new materials), or on the contrary, does the key to success for new profit strategies lie in the social uses of the automobile (urban vehicles with multiple ownership, widespread rental, etc.)*?" A decade on, the car industry is having to face all of these challenges and choose an "optimal" position. This involves at once seeking and achieving maximum volume, the required diversity, quality and innovation, and permanently reducing costs at constant volume.

For Womack et al. (1991), "*The future is already written. The only survivors will be those with the courage to adapt and apply the best solutions, resulting from 'lean production*'". The debate remains open on the strategies to apply in different regions of the world, faced with the growing need for mobility and an obligation to respect the environment. The future also involves meeting a rising demand for communicating vehicles (Xerfi Study on the 2020 horizon, 8 % of future buyers). Using embedded systems and communication technologies, vehicles can communicate with each other and with the road infrastructure via the internet and local networks, bringing users access to new safety, intelligent navigation and personalization features.

Cars will need to complement other transport solutions in innovative production models that remain to be defined. Making these models a reality will mean rethinking the way that the innovation process is organized so that the chosen production model fits in with mobility requirements, i.e. it needs to be customer-focused (Gatignon and Xuereb 1997; Gotteland et al. 2007).

2.3.3 Public-Private Partnerships for an Overall Approach to Mobility

Yet, can state action create the conditions for new production models to emerge? Another issue raised by the current ecosystem is the connection between public policies and major innovation opportunities, such as electric or autonomous vehicles. Have the constraints of public policies aimed at reducing CO_2 emissions in fact created a *strategic opportunity* for car manufacturers? In other words, have manufacturers transformed an environmental constraint into a development opportunity, for example by reconfiguring their value chain or designing new products and services? It is clear that these regulations have put significant constraints on manufacturers, forcing them to redefine their value chain, alliances and business models.

Several recent examples illustrate this trend, with new programs emerging between different stakeholders to produce cleaner, more energy-efficient and autonomous cars. Thus, the European Union project Elibama gathers car manufacturers, battery suppliers, recyclers and universities; another project led by PSA-Peugeot-Citroën, AbattReLife, is built on the same model; and private-public partnerships like the EGVI-EU project aim to accelerate R&D on clean technologies.

These projects show that various partners are keen either to maintain a place in this new ecosystem, which is the case for manufacturers, or to enter it and position themselves as "key players" in the new value chain, which is the case for battery suppliers, recyclers and high-tech companies. However, it is worth mentioning that these projects or research programs are actively supported by public authorities since they play a major role in developing the new mobility ecosystem. They organize mobility by setting up services (charging stations, self-service cars) and provide the necessary impetus for developing alternative technologies, creating de facto alliances between car manufacturers, car park operators, spare parts manufacturers, insurance companies and user representatives.

Although the targets are clear, one question remains, which is that of the overall governance required to coordinate all of these decision-making levels. Stakeholders are increasingly numerous and diverse, regulations are put into question, and local authorities are confronted with new modes of transport that oblige them to reorganize the way that their territory is managed. Urban policies are challenged and obliged to redefine their investment choices and set up incentive measures or usage restrictions in urban areas.

2.3.4 The Customer, the "Kingpin" of Expected Value Creation?

Another issue that should be considered by manufacturers and other stakeholders in the industry is the new requirements and usages of customers, who, from one generation to the next, confer cars with a different role and social status. Generation Y, which emerged over ten years ago, has introduced new societal values, including a very different relationship to mobility. This generation designates 25–35 year-olds, also known as "digital natives" and "Generation Why", who grew up within the internet ecosystem and have built a new generational identity through social networks. Numerous surveys have analyzed their behavior, preferences and lifestyle (Dagnaud 2013).

These young people, the successors of Generation X (Coupland 1991), live in a system of constant communication dominated by new usage types; they expect to use objects and products for free or almost nothing, enjoy open access to knowledge technologies and, in particular, consider that ownership can be shared or exchanged and is not a symbol of social success. This is particularly striking for example in their representation of the car (Parasuraman et al. 2013). Lastly, the idea of a "*reversibility between producers and consumers*" is interesting, and has been the subject of several original research papers. This situation occurs, for example, when the owner of an electric car produces her own electricity to drive and may even sell it to a third party to make a profit! (Kempton and Tomić 2005).

A study commissioned by Deloitte consulting company in 2014 on Generation Y's relationship with mobility illustrates our point.[2] The survey questioned 23,000 young people from around twenty countries in North America, Europe and Asia. When asked to rank criteria for the future purchase of a car, the young people put cost and utility first, followed by ecology and pleasure, and lastly technology and safety. What is striking about these choices is the low-cost requirement for purchasing a vehicle, which is very different from past generations, who placed social image and status recognition above cost, even if it meant making financial sacrifices.

The second interesting point is the ecological requirement, which implies that this generation would happily accept a move from internal combustion cars to electric cars. In their replies to questions on vehicle ownership, these young people clearly positioned themselves as "*very favorable to alternative solutions: car sharing, carpooling and rental*". Their priority is optimal, "comfortable" journey solutions, the lowest cost for a means of travel that significantly saves time. This also involves benefiting from using mobility without the concern of managing it.

Lastly, young people replied favorably to a question on advanced automation and future driverless cars, in particular stating that it is, "*difficult to stop using technologies when I'm driving, I want to stay connected to my friends and family, I want to carry out tasks while I drive*". To sum up, in terms of the connection between this generation and mobility, Generation Y members have mobility requirements that are very close to their ideals, preferring a community of users to private, personal transport; they want freedom to travel at a low cost, want to live spontaneously, and especially stay permanently connected to social networks. Their lifestyle reflects their mobility choices: to connect with their "tribe", be informed of everything immediately, and be able to do several things at the same time. The connected/autonomous car is clearly the solution that responds to these new requirements.

However, the question of whether this new economic and ecological model would be acceptable to customers is not so simple (Bühne et al. 2015) for a number

[2]Study carried out in 2014 by students from ESSEC, ECP and Dauphine universities for Deloitte Conseil on Generation Y and Electromobility.

of reasons. Consumers will soon be able to choose between different means of travel: connected cars, autonomous cars, robot taxis, shared shuttles, etc. But will users be capable of defining their own strategy to match their needs?

In addition, a vehicle's energy efficiency cannot be dissociated from its price, which must remain attractive. The question of the cost of using a vehicle raises another, more complex issue, i.e. how the vehicle's *energy source* is managed. Users must, for example, adapt to the new services on offer for charging their vehicle with electricity. Creating efficient charging facilities in a town requires a densely meshed electric grid with significant power.

This type of innovation transforms the traditional interface between cars and users and, at the same time, creates a structural, systemic interaction between towns, energy and information. That means a radically new business model.

2.4 Conclusion

The solutions to this profound change in mobility come from both private initiatives (carpooling, car sharing) and public initiatives through tax incentives, regulations, and investment in infrastructures to encourage intermodality. The economic, social and technological consequences are considerable, since they modify the place of traditional stakeholders in the automotive industry (car manufacturers, spare parts manufacturers, suppliers, service operators, recyclers, etc.) and raise the issue of organizing, managing and sharing urban space. Cooperation, alliances and partnerships are taking shape around the world involving all car manufacturers, since the question of leadership will be crucial in determining who will stay in the race and who will drop out.

In this new mobility paradigm that we have called *auto-mobility*, car manufacturers are subject to greater complexity and uncertainty when making their strategic decisions. In this respect, technological questions are correlated to societal issues, including whether users accept a new mobility model that must first satisfy their needs.

The traditional concept of a powerful, personal car is being replaced by other types of mobility. The social symbols, priorities and preferences of young people are no longer the same. Real-time access to information has become essential. Mobility is more important than power. People want to remain constantly connected to their environment. Cars can no longer be considered as an end in themselves, but as tools for gaining mobility. These new services contribute to the arrival of new types of mobility that are more relaxed, rational and sustainable, and to the emergence of a new ecosystem.

Acknowledgment This research benefits from the support of the Chair "PSA Peugeot Citroen Automobile: Hybrid technologies and Economy of Electro mobility", The Armand Peugeot Chair led by CentraleSupélec, ESSEC and sponsored by PEUGEOT CITROEN Automobile.

References

ADEME, 2012. Contribution de l'ADEME à l'élaboration de visions énergétiques 2030-2050. Agence de l'Environnement et de la Maîtrise de l'Energie.

Arjaliès, D.L., Goubet, C., Ponssard, J.-P., 2011. Approches stratégiques des émissions CO2. Revue française de gestion 123–146.

Arjaliès, D.-L., Ponssard, J.P., 2010. A Managerial Perspective on the Porter Hypothesis-The Case of CO2 Emissions. Palaiseau, Éditions de l'École Polytechnique, Crifo P., Ponssard J.-P., eds 151–168.

Artus, P., Mistral, J., Plagnol, V., 2011. L'émergence de la Chine: impact économique et implications de politique économique. Paris: Conseil d'Analyse Économique.

Attias, D., 2013. Comment concevoir le véhicule de demain? Comment en parler? Audition publique de l'OPECST, Les nouvelles mobilités sereines et durables: concevoir des véhicules écologiques.

BlaBlaCar, 2016. . BlaBlaCar. URL https://www.blablacar.fr/ (accessed 6.4.16).

Boyer, R., Freyssenet, M., 2006. Le Monde qui a changé la machine. Essai d'interprétation d'un siècle d'histoire automobile. Quatorze textes préparatoires. GERPISA, Paris. Éditions numériques: http://freyssenet.com.

Bühne, J.-A., Gruschwitz, D., Hölscher, J., Klötzke, M., Kugler, U., Schimeczek, C., 2015. How to promote electromobility for European car drivers? Obstacles to overcome for a broad market penetration. Eur. Transp. Res. Rev. 7, 1–9.

CGDD, DGE, PIPAME, 2016. Usages novateurs de la voiture et nouvelles mobilités. French Ministry for Ecology, Sustainable Development and energy.

Coupland, D., 1991. Generation X: Tales for an Accelerated Culture. St. Martin's Griffin, New York.

Dagnaud, M., 2013. Génération Y: Les jeunes et les réseaux sociaux, de la dérision à la subversion, 2° édition. ed. Presses de Sciences Po, Paris.

Deblock, C., 2012. Présentation du dossier : Innovation et développement chez Schumpeter. Revue Interventions économiques. Papers in Political Economy.

Freyssenet, M., 2015. La production automobile mondiale, des quatre continents et des principaux pays constructeurs, 1898-2015. Document d'enquête: quatre tableaux et six graphiques commentés. Édition numérique, freyssenet. com, 2007 et mises à jour.

Freyssenet, M., 2011. Three possible scenarios for cleaner automobiles. International journal of automotive technology and management 11, 300–311.

Freyssenet, M., 2009. Sommes-nous au début d'une seconde révolution automobile ? Presented at the 176ème journée du GERPISA, CCFA, Paris.

Frigant, V., Jullien, B., 2014. Comment la production modulaire transforme l'industrie automobile. Revue d'économie industrielle 11–44.

Gatignon, H., Xuereb, J.-M., 1997. Strategic orientation of the firm and new product performance. Journal of marketing research 77–90.

Gotteland, D., Haon, C., Gauthier, C., 2007. L'orientation marché: synthèse et nouvelles directions théoriques. Recherche et Applications en Marketing 22, 45–59.

IEA, 2012. Energy Technology Perspectives 2012: Pathways to a Clean Energy System. International Energy Agency, Paris.

Jullien, B., Pardi, T., 2013. Structuring new automotive industries, restructuring old automotive industries and the new geopolitics of the global automotive sector. International Journal of Automotive Technology and Management 20 13, 96–113.

Kempton, W., Tomić, J., 2005. Vehicle-to-grid power fundamentals: Calculating capacity and net revenue. Journal of power sources 144, 268–279.

Klappenecker, A., Lee, H., Welch, J.L., 2014. Finding available parking spaces made easy. Ad Hoc Networks, (1) SCEnarios for ad hoc Network Evaluation Studies (SCENES); (2) Modeling, Analysis and Simulation of Wireless and Mobile Systems & (3) Algorithmic Aspects of Ad-Hoc Networks 12, 243–249.

KPMG, 2015. KPMG's Global Automotive Executive Survey 2015.
Nazem, M., Trépanier, M., Morency, C., 2011. Demographic Analysis of Route Choice for Public Transit. Transportation Research Record: Journal of the Transportation Research Board 2217, 71–78.
Observatoire Cetelem, 2015. Les marchés automobiles dans le monde.
OECD, 2013. Industrie automobile : ajustements à moyen terme des capacités de production.
Oliver Wyman, 2015. Oliver Wyman Study on Automotive Sales of the Future- System Profit 2035: Automakers Need to Reinvent Sales System.
Parasuraman, A., Hoefnagels, A., Migchels, N., Kabadayi, S., Gruber, T., Loureiro, Y.K., Solnet, D., 2013. Understanding Generation Y and their use of social media: a review and research agenda.
Plassat, G., 2012. Mobilités et Transports du Futur - Quelles rencontres pour le 21ème siècle ?
PWC, 2014. Autofacts, Study of global vehicles production.
Qian, L., Soopramanien, D., 2015. Incorporating heterogeneity to forecast the demand of new products in emerging markets: Green cars in China. Technological Forecasting and Social Change 91, 33–46.
Schaefers, T., 2013. Exploring carsharing usage motives: A hierarchical means-end chain analysis. Transportation Research Part A: Policy and Practice 47, 69–77.
Womack, J.P., Jones, D.T., Roos, D., 1991. The machine that changed the world: the story of lean production. Harper Collins.

Chapter 3
The New Mobility Paradigm. Transformation of Value Chain and Value Proposition Through Innovations

Guy Fournier

Abstract Meeting the need for transportation when the population will reach nine billion people in 2050 will be challenging for our societies: Driven by external costs like global warming, noise or congestions and supported by new policies and growing customer awareness, the current mobility paradigm based on cheap fossil fuel energy and high CO_2 exhausts comes to its social, economic and environmental limits. Innovation can provide value propositions to meet the mobility needs of future generation. Innovation in products will foster new energy efficient, low CO_2 emitting electric vehicles. Innovation in services, triggered by the net economy, will simplify travels, improve the value of time, the use of assets like cars, bikes, parking, taxis, plane etc. while driving or parking and emphasise environmental aspects. Customers can in this way cut costs, extend the mobility means, meet people etc. Bundling mobility solutions will further facilitate access to seamless mobility experience. Jugaad innovation and reverse innovation finally can meet the needs of developing countries and subsequently of developed countries and provide methods on how to do more with less. The value propositions of mobility solutions will therefore deeply impact the future: new raw materials, components, vehicles and services will emerge; new players will reshape the value chain, capture competitors' customers and customer value, thus challenging traditional OEM's with new products and services; even customer will be part of the value chain and become prosumer (either consumer or producer, as the case may be). Simultaneously sustainable mobility will rise and propose an answer to the current systemic challenges and externalities our societies are facing.

Keywords Sustainable mobility · Future mobility · Low carbon economy · Electric vehicles · Powertrain · Sharing economy · Multimodal mobility · Mobility service · Servitization · Jugaad innovation · Reverse innovation · Circular economy

---

G. Fournier (✉)
Institute for Industrial Ecology, Pforzheim University of Applied Sciences, Tiefenbronner Straße 65, 75175 Pforzheim, Germany
e-mail: guy.fournier@hs-pforzheim.de

D. Attias (ed.), *The Automobile Revolution*, DOI 10.1007/978-3-319-45838-0_3

3.1 Introduction

In the fifties there were approximately 50 million vehicles in the world. In 2014 there were nearly 900 million and this number is expected to rise to 1.3 billion by 2030 and 2 billion by 2050. The International Monetary Fund (IMF) expects even 2.9 billion cars worldwide by 2050 (Chamon et al. 2008). Currently, in the developed world, all the personal mobility needs are essentially provided by light-duty vehicles which are fuelled by fossil energy. The diffusion of vehicles is expanding throughout most of the developing countries due to growing population coupled with rising incomes (Wolrd Business Council for Sustainable Developement 2004). This raises the question of how we can cope with distance in the future when the population will reach nine billion people (International Transport Forum 2011; World Energy Council 2011).

To produce and drive these light-duty vehicles and to satisfy the growing mobility needs, additional resources are necessary and external costs due to rising urbanisation, pollution, accidents or global warming will be generated. Thus, it becomes more and more urgent to question if our mobility paradigm will reach the limits of our natural, economic and social systems.

The aim of this chapter is to examine, with an integrated interdisciplinary approach, why the changes mentioned above will affect individual mobility, and if innovation can provide an answer to these challenges. In a first step, the limits of the current mobility paradigm will be sketched. Afterwards, innovation in individual mobility and especially new biofuels and new powertrains, like electric vehicles (EV), fuel cell vehicles (FCV) or other technologies, will be examined and evaluated. Then innovation in mobility services and value added services will be analysed as a tool to increase the productive use of assets and make mobility more affordable and more customer oriented. Finally, it will be discussed if innovation in developing countries and especially jugaad innovation (see Sect. 3.5) can help us face the mentioned challenges and develop a new sustainable mobility paradigm.

3.2 Limits of the Current Mobility Paradigm and Drivers for Change

The current mobility paradigm is based on fossil fuel and individual mobility. This paradigm will reach environmental, economic and social limits (see Sect. 3.2.1), thus provoking a change of the political frameworks (see Sect. 3.2.2) and, together with technological progress (see Sect. 3.2.3), driving the change to a more sustainable mobility (see Sect. 3.2.4). Figure 3.1 states the process (see Fig. 3.1).

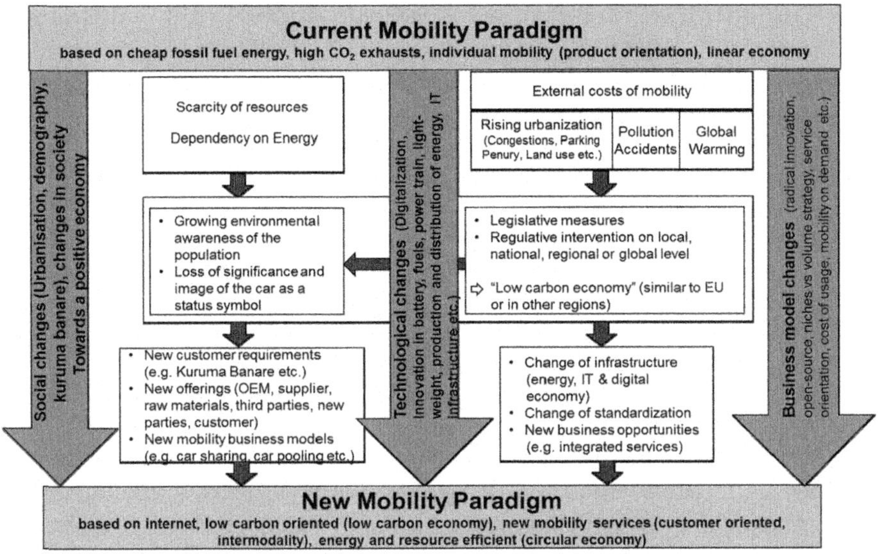

Fig. 3.1 Drivers of the new mobility paradigm. *Source* Based on Fournier et al. (2012)

3.2.1 Ecological, Economic and Social Limits

The limits of the current paradigm can be seen first of all in the availability of energy and resources to produce and drive two billion or more vehicles. Today, individual transportation is predominantly enabled by fossil fuels. However, the global oil reserves are limited and in the same time the combustion of fuel is harmful to nature and environment (Fournier and Stinson 2011).

Due to growing vehicle fleets and vehicle ownership (vehicles per 1000 people), it is expected that the demand for liquids for transport will increase steadily from 50 to nearly 65 mb/d (millions of barrels per day) by 2035, driven essentially by non OECD (Organisation for Economic Co-operation and Development) countries (BP 2016). The Energy Information Administration (EIA) has estimated a worldwide consumption of 75 mb/d by 2040. On the other side, the delivery rate of conventional crude oil cannot be significantly increased anymore. It might even be the case, that the production maximum has been already reached (Energy Watch Group 2007). The gap between demand and supply can only be closed by non-conventional oils and gases, while not yet discovered oil reserves need to be found and exploited additionally (IEA 2010). Recently innovation in technology and huge gains in productivity have in fact unlocked vast resources of tight oil and shale gas (BP 2016). Therefore, in 2014 the USA became the world largest oil producer ahead of Saudi Arabia (BP 2015). However, closing this gap between offer and growing demand will increase the cost of extraction as depleting onshore fields in the Middle East will be substituted by more costly deep water, ultra-deep water, shale or sand based oil extraction.[1] The reason is the curve of world marginal cost of extraction, showing, by type of oil extraction, how much it costs to produce an additional barrel. Thus, due to the evolution of supply and the increasing cost of extraction combined with the expected strong rise of oil demand the International Energy Agency (IEA) and the EIA predict increasing costs of liquids (IEA 2010; EIA 2016).

In addition to this economic motivation, reducing the environmental negative effect generated by the production and use of oil (especially non-conventional oils) is another key driver of the new mobility paradigm. The accumulation of carbon dioxide together with other greenhouse gases in the atmosphere is strongly suspected to be the main reason for the increase of the average global temperature within the last 50 years (Intergovernmental Panel on Climate Change 2007, 2014). Studies of IEA and World Wildlife Found (WWF) confirm this fact and warn of the consequences of global warming (Fournier and Stinson 2011).

Beside the problem of global warming, transport causes other external costs: in Germany the external costs of passenger transport in 2000 (including global warming, accidents, air pollution, noise, etc. and excluding costs of traffic congestion) were estimated at 67 €/1000/pkm (passenger kilometre). At European level, the costs of passenger transport are estimated at 1.1 % of the GDP (100 bn €) for global warming, noise and air pollution and another 1.1 % of the GDP (100 bn €) for traffic congestion (Maibach et al. 2008). Moreover, in some American cities more than a third of all the land is now dedicated to parking (Kalanick 2016). During the last years more detailed studies have been published about air pollution. In France the costs of air pollution have been estimated at 68–97 bn € per year (2010). Considering especially the particulate emissions, the costs are valued between 20 and 30 bn €, which corresponds to approximately 400–500 € per inhabitant (Senat: commission d'enquête 2015). The OECD estimated furthermore that costs of the health impact of outdoor air pollution from road transport in OECD countries account for 1 trillion USD in 2010. The costs in developing countries like China or India are probably higher (OECD 2014).

Due to our fossil based economies, the discussion today is essentially focussed on the availability of fossil fuels, global warming or air pollution. These are important aspects but not the only ones: in fact our world is facing a problem of limited availability of easily accessible materials and energy (Ellen Macarthur Foundation 2013, 2014) and a problem of disposal on a more general level.[2] The current economies have been linear since the industrial revolution with take, make, and dispose as a basis (Ellen Macarthur Foundation 2014). With the upcoming expected billions of consumers from the developing world, which will enter the middle class and adopt the western consumption habits, this linear system comes to its limits. This is also true for the aforementioned car market perspectives.

[1]See details about the global liquid cost curve in USD/oil barre: (Nlysveen 2015).

[2]This discussion started with Malthus (1798), Ricardo (1822) and more recently Club of Rome (1974).

A circular economy could by contrast decouple growth from resource constraints, avoid waste and pollution, and preserve and enhance natural capital. Copying biological cycles, technical cycles have to collect, maintain, prolong, reuse, redistribute, refurbish, remanufacture and recycle products and parts to minimise systematic leakage and negative externalities (Ellen MacArthur Foundation 2015).

Supplementing the above investigation of economic and environmental issues, the following section addresses from a more general perspective the social question since economic actors are more and more aware of the limits of our systems.

When observing current tendencies in societal developments, on the one hand an increasing urbanization can be identified. According to the United Nations more than 50 % of the world's population was living in conurbations for the first time in 2009 and it is estimated that this number will increase up to 70 % until 2050 (United Nations 2010). Increased traffic intensity, local air pollution and the lack of parking facilities are problems to be expected within this development.

On the other hand, there are changes of the customer requirements in respect of the future mobility which will influence market opportunities. The privately owned car will be more and more challenged as a solution to satisfy mobility needs in crammed urban areas with good public transport and more and more restrictive city regulation and taxes for vehicles. At the same time the needs to stay mobile remain on a high level (Teichmann 2011). As a result, young people show a decreasing interest in the automobile. This was observed for the first time at the beginning of the 90s in Japan and was described with the term "Kuruma Banare". 'Kuruma Banare' may be translated by the English term 'demotorisation' describing an attitude towards mobility without personally owning a motor resp. automobile. Statistics made by the German Federal Motor Transport Authority also showed that in Germany, in 2009 the proportion of customers in the 18–29 age segment buying new vehicles dropped by 50 % within the last ten years as compared to 1999 (Dalan and Dol 2010). In the USA the number of persons aged 18–39 with a driving license declines in a similar way (Schoettle and Sivak 2014). In France the same phenomenon can be observed: the lack of interest in buying a vehicle is explained by the price of the vehicles, the absence or delay of their connectivity and the lack of alternatives like car-pooling (Dupond-Calbo 2015).

These social changes indicate that through new approaches in the construction and development of vehicles as well as in the design of mobility concepts, new market opportunities in the automotive market might arise.

3.2.2 Political and Legal Drivers for a New Mobility Paradigm

Together with the enforced dependency on oil suppliers and oil exporting countries, negative effects on economic growth as well as environmental degradation are expected if no effective measures are taken to increase efficiency and promote alternatives to the current mobility paradigm.

Governments and politics react to the above mentioned issues in many respects. Regulatory interventions on the international (e.g. post-Kyoto protocol agreement, “low carbon economy”-strategy of the European Union), national (e.g. bonus-malus system in France) and local (e.g. inner city tolls in London) level can be identified.

With respect to global warming, UN Secretary-General Ban Ki-moon demands from all involved parties—governments around the globe as well as top managers of key players in industry—to re-engineer the current techno-economic paradigm and implement all necessary political measures as early as possible (Ki-moon 2011). The Paris Agreement (“accord de Paris” in French) at the Conference of the Parties on Climate Change 2015 dealt with the mitigation of greenhouse gas emissions; adaptation and finance will start in the year 2020. The agreement was adopted by consensus by all of the 195 UNFCCC (United Nations Framework Convention on Climate Change) participating members states and the European Union (EU) (United Nation 2015). The EU has committed to further reduce the greenhouse gas emissions until 2020 by at least 20 % compared to the status quo in 1990. This resulted in the regulation from 2009, which aims at reducing the CO_2 emissions of passenger cars, responsible for 26 % of the CO_2 emissions within the EU (Bundesministerium für Umwelt, Naturschutz, Bau und Reaktorsicherheit, BMUB 2009). This regulation gives a legal framework regarding the limitation of future CO_2 emissions admitted to the European automobile industry. It includes the binding gradual reduction of CO_2 emissions of the fleets of automotive OEMs until 2020 to 95 g CO_2/km. However, it is likely that these restrictions, perhaps in combination with subsidies and other measures, will be enforced for future decades. Speculations even reach to a restriction between 10 g and 35 g CO_2/km for the worldwide automobile industry by 2040 to achieve global warming below 2 °C until 2100 (Unknown 2010).

Politics are increasingly enforcing the concept of a circular economy: in Europe a directive on end-life vehicles has been adopted (Directive 2000/53/EC of the European Parliament and of the Council 2000). In France, within the framework of the “Energy Transition for Green Growth Act”, waste is tackled and circular economy promoted (LOI n° 2015-992 2015). As a result a decree has been published in 2016 making it obligatory for automobile garages to provide used spare parts from 2017 on (Décret n° 2016-703 2016). This could be beneficial for the customer in terms of lower prices of the parts. It also has environmental benefits through reduced material consumption and CO_2 emissions and preservation or creation of jobs that can be difficult to relocate out of the country (Unknown 2014).

These restrictions can be seen as additional and accelerating factors driving the change of the current mobility paradigm.

3.2.3 Technological Drivers for a New Mobility Paradigm

Innovation in vehicles can drive the change for individual mobility. As a first step numerous vehicle manufacturers are optimizing the combustion engine in order to reduce the fuel consumption and thereby the correlating CO_2 emissions: cylinder deactivation, variable valve train, turbocharging and downsizing, utilization of exhaust gas energy, direct injection, new combustion, variable compression etc. can improve the current powertrain technologies (Wallentowitz and Freialdenhoven 2011).[3] Further possible strategies are the improvement of efficiency through aerodynamics, improved drive resistance, improved energy efficiency of car components (e.g. power steering, air conditioning, alternator) or light weight design. Another approach could finally be to look for alternative fuels like biofuels, hydrogen or electricity. A brief evaluation in terms of CO_2 emissions and energy efficiency of the mentioned alternative fuels will be conducted within the next section.[4] These incremental and radical innovations focus on improving the current mobility paradigm but do not impact the mobility paradigm as a whole.[5]

Digitalisation is another driver of change which will transform our economies by shaping the 4th industrial revolution (Schwab 2016). Increasing returns (economies of scale, economies of scope), network effects and lock-in effects are the keys of this new economy (Fournier and Donada 2016)[6] which will reshape the industry structure (Porter and Heppelmann 2014). The current auto-industry's value chain structure and internal rivalry between OEMs will evolve towards a play of competing mobility eco-systems with new parties and with even the customer as a mobility partner (Fournier and Donada 2016). The customer and the satisfaction of his/her mobility needs will move in the focus of interest (customer centricity) and generate the necessity of a holistic approach in which products and services are

---

[3]See for the impact of these technologies for consumption reduction: p. 40; pp. 71–89.

[4]See for more details: Fournier and Stinson (2011).

[5]Four categories of innovations have been identified by Freeman and Perez: incremental innovations, radical innovations, new technological systems (systemic innovations), and technological revolutions or new techno-economic paradigms. New techno-economic paradigms represent changes in technological systems that are so far-reaching in their effects that they have a major influence on the behaviour of the entire economy. See Freeman and Perez (1988).

[6]In contrast to traditional goods, digital goods have declining marginal costs (Clement and Schreiber 2013; Rifkin 2014). Marginal costs of digital goods tend to zero, because copying digital goods causes almost no costs (Varian 2000) and fix costs may be lower. This is linked with the indivisibility and appropriability properties of information. Economies of scope describe efficiencies wrought by variety, not volume. They occur in the net economy through cross selling e.g., through referral systems or links which connect own products and services with products and services from partners or vice versa. Since the marginal cost of adding an extra commodity is negligible, selling two or more distinct goods together for a single price can be attractive as well. The benefits of a product depend on the number of users: In the past the value of a commodity depended on the price and quality. Within a network, not the scarcity of a commodity, but the degree of distribution determines the value of a commodity. This phenomenon is called network effects.

bundled and offered to consumers as a system. Through digitalisation and especially reduction of transaction costs, new leaders of the value chain (shapers) will emerge to satisfy these needs. The shapers will be able to combine bicycles, scooters, cars or trains into a one face to the customer mobility service and trigger new and innovative business models with an added value for the customer. To provide these services the shapers will need to associate partners from the automotive industry, information technology, telecommunications, utilities and mobility service providers such as car sharing or railway companies (Fournier and Donada 2016). The process of growing digitalisation and servitization, which disrupts established economic and business rules of thumb and creates a "creative destruction" process (Christensen 1997), will be detailed in Sect. 3.4.

3.2.4 The New Mobility Paradigm

To summarize, ecological, economic and social drivers as well as political and legal influences will trigger the change of the current mobility paradigm towards a mobility paradigm based on low emissions of greenhouse gases and an efficient usage of energy resources to provide individual mobility related with new, not ownership-oriented but need-driven, mobility services.

On a more abstract level, the changes can be understood as a consequence of the change of a techno-economic paradigm. The division of labour and the resulting growth of specialization can be seen as our source of wealth (Smith 1776). This economic system like all systems in our universe necessitates energy, information and materials (Durand 1993). If the purchase (scarcity of oil or other substitutes) or the use (CO_2 restrictions against global warming) of energy expand further, the division of labour will fundamentally change at the company level and at the local, national, regional and global level. If the digitalisation of our economy lowers the cost of transaction, the division of labour will be profoundly impacted as well: Information is necessary to coordinate the division of labour within a company, between companies and between companies and customers (for instance through internet). If the availability of natural resources and materials causes a problem, an issue is to the current linear economy towards a circular economy.

The creation ("creative destruction") (Schumpeter 1950) and development of new products, new industries, services and infrastructure (Perez 1983) will impact especially the current mobility paradigm based on cheap fossil fuel energy and high CO_2 exhausts and therefore foster the emergence of a low carbon oriented, energy efficient mobility paradigm (see Fig. 3.1). The possibilities to achieve this transformation depends of course on social behaviours. Even though other less optimistic scenarios are possible,[7] this disruptive scenario described above seems to be the most feasible and probable if the challenges our societies are facing should be met.

[7]For more details about the conditions of a new automobile revolution see Freyssenet (2009).

3.3 Innovation in Individual Mobility to Meet the Challenge: New Powertrains

Following this new mobility paradigm new solutions and value propositions for the future individual mobility become indispensable. In order to achieve the next steps into that direction, alternatives are to be created, how automobiles will be constructed and how mobility will be shaped in the future.

3.3.1 Towards Electrification of Powertrains

The first recognizable step, numerous vehicle manufactures were and are taking, is the ongoing optimization of the combustion engine in order to reduce fuel consumption and thereby the correlating CO_2 emissions. But still the concept of the combustion engine—even though characterized by good performance, cruising range, low costs, and driving comfort—has its shortcomings: e.g. depending on the type of engine, it shows more than 70 % of thermal losses and just 15–20 % effective energy to drive compared to the primary energy applied. These are natural limits of technological performance which cannot be improved significantly since 1911 where already 12.5 % effective energy was tested (Fournier and Stinson 2011).

Biofuels were seen for a long time as a promising alternative to decrease CO_2 emissions but are criticised more and more. The first generation of biofuels is based on food crop like colza, wheat or sugar cane. The 2nd generation of biofuels are derived from plants and include all the biomass—stems and leaves—or waste. The 3rd generation of biofuels is based on microalgae (see Fig. 3.3).[8] New well-to-wheel calculation taking the direct emissions from cultivation, processing, transport, and distribution and the emissions from land use change showed that 1 litre biodiesel increases transport GHG emissions by 80 % in comparison with fossil diesel, while 1 litre bioethanol emits 30 % less than fossil petrol. Advanced non-food based biofuels by contrast can save 17 % CO_2 (Globim 2016). Another problem is that these substitutes for fossil fuels will not be competitive before 2030. In addition, disadvantages such as competition for agricultural land along with rising nutritional requirements, potential damage to biological diversity as well as social implications should be considered (IEA 2011).

A further step of development is the electrification of the powertrain. Hybrid vehicles using a subordinate electric motor in addition to a combustion engine are grouped into the four levels micro, mild and full hybrid as well as plug-in hybrid.

The advantages in fuel consumption range from 7 to 20 % for micro and mild-hybrid vehicles, which only have either a start/stop function or a small electric

[8]A detailed evaluation of biofuels can be found at Roland Berger Strategy Consultants (2016).

support engine Wallentowitz and Freialdenhoven (2011) to 30–40 % for full hybrid concepts, which allow purely electrical driving over short distances (Stan 2008). Due to the design of the plug-in hybrid vehicle, where the combustion engine is only the back-up solution for longer distances, which cannot be driven purely electrical, these vehicles offer the greatest potential for reduction in fuel consumption (Wallentowitz and Freialdenhoven 2011) within the group of hybrid technologies. However, this depends highly on the capacity of the installed battery and customer behaviour.

The next step in the electrification of the powertrain is the pure electric vehicle, allowing permanently emission-free mobility as long as renewable energies such as solar or wind energy are used (Wüchner 2007). Electric cars, which gain the energy for the drive of the electric motor from battery, currently offer the largest potential in efficiency regarding the type of primary energy applied (Wallentowitz and Freialdenhoven 2011).

3.3.2 Comparison and Interpretation of Different Well-to-Wheel Scenarios

Drawing the bottom line from today's perspective, it is unlikely to achieve the EU-target of 95 g CO_2/km until 2020 exclusively by the optimization of combustion engines or hybrid vehicles. By 2017 the new world standard WLTP (Worldwide harmonized Light Vehicles Test Procedures) will take the place of the updated and less rigorous NEDC (New European Driving Cycle). The target will then be more difficult to reach. In order to even obey the predicted limit of 10–35 g CO_2/km in a country like Germany, a rate of at least 70 % CO_2-emission-free driving has to be achieved until 2050. Without enhancing the efficiency of combustion engines this figure is expected to rise even up to 83 % (Unknown 2009). Therefore electrical mobility is indispensable to achieve this goal and keep global warming below 2°.

Figure 3.2 draws and compares the different alternative technologies regarding efficiency and emissions of energy used according to a Well-to-Wheel balance. It can be seen that battery-electric vehicles (BEV) based on nearly emission free renewable energies have high energy efficiency and low emissions.

Besides, Fig. 3.3 shows the superiority of electric vehicles in comparison to biofuel-powered vehicles with combustion engines. Thus on one hectare land it is much more effective to produce electric energy from renewable energies than planting agricultural plants to produce biofuels to achieve greatest possible range (Fournier 2009; Fournier and Stinson 2011).

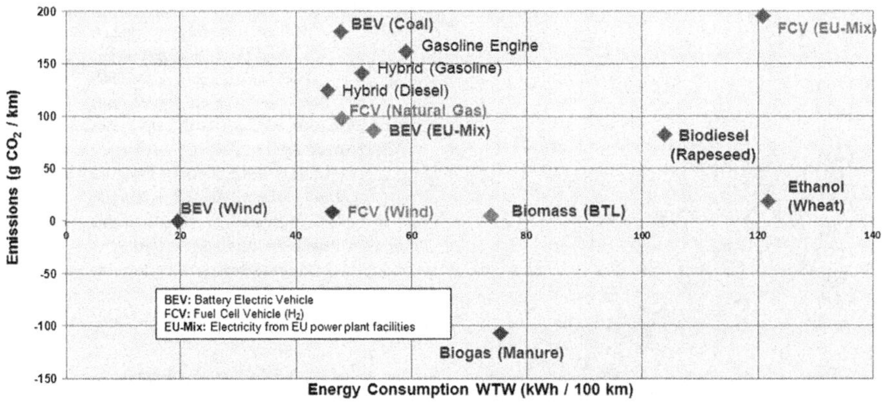

Fig. 3.2 Well-to-Wheel efficiency of individual transport: CO_2-emissions and energy demand. *Source* Fournier et al. (2012) based on Concawe et al. (2007), Optiresource-Software (n.d.)

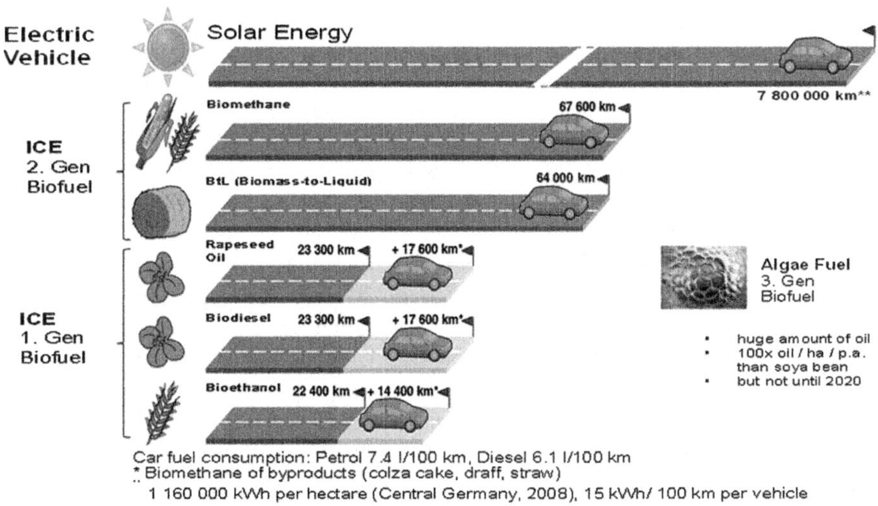

Fig. 3.3 Cruising range to be achieved with one hectare land. *Source* Fournier (2009, 2012)

3.3.3 Diffusion of New Powertrains and Limits

The market of EV is actually growing very fast: China is in 2015 the biggest (38 % of global market sales) and fastest growing market (+245 % between 2014 and 2015); Western Europe is the second biggest market with 33 % of global market sales; US market represents 20 % of global sales (Avere-France 2016). But global automobile production is 90.68 Million in 2015. The EV market is thus lower than 1 % of the global market. Despite the efficiency and emission advantages of electric vehicles, the EV markets remains very small and dependent on government

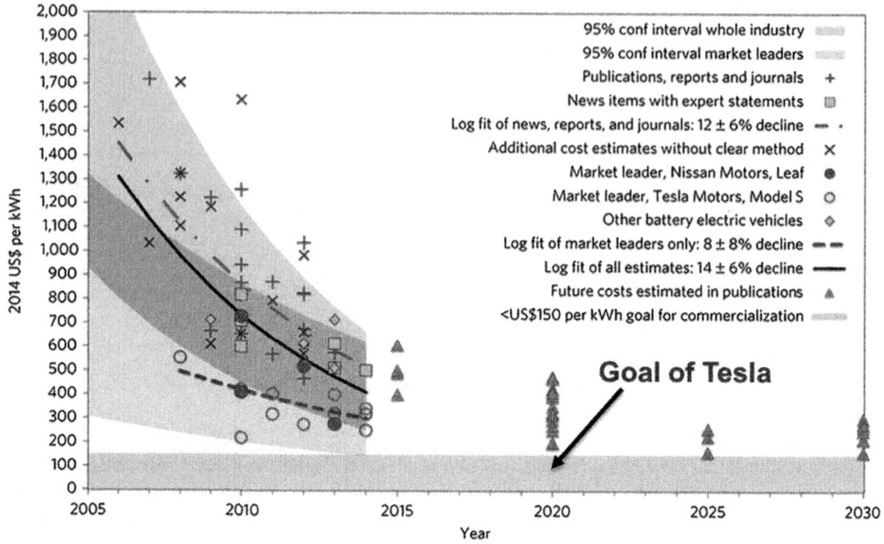

Fig. 3.4 Battery: key success factors for EV and Hybrid. *Source* Nykvist and Nilsson (2015)

policies: battery costs, the comparatively high weight as well as the small energy and power density of the battery (leading to low ranges) and finally the hardly existing charging infrastructure are the major obstacles to a faster adoption and diffusion of EV (Fournier and Stinson 2011).

The evolution of battery costs shows in fact that the cost in US $/kWh (kilowatt-hour) dropped from 1300 $/kWh in 2006 to 450 in 2014 but are still above 150 $/kWh, which is the limit for commercialisation and large diffusion of EV's (see Fig. 3.4). According to this analysis, which summarises multiple type of source, the price to enable a large diffusion is expected by approximately 2025 (Nykvist and Nilsson 2015). On the other side, the goal for Tesla, an American start-up dedicated to the construction of EV's, has the target to produce batteries bellow 100 $/kWh by 2020 (OECD/IEA 2016) with the help of a Giga battery factory. According to Tesla, their forecast is to produce 500,000 eV's by 2020 and use economies of scale to lower the costs of batteries. The models of Tesla are today furthermore very promising as in just a few weeks the newest model was pre-ordered 400,000 times worldwide 1000 $ of reservation deposit included (Parin 2016). Tesla is now leader with the EV model S in USA and Germany in comparison with ICE powered class S from Daimler (Spath 2016).

Though it is difficult to predict the exact timing of the shift from fuel powered cars to EVs, mass adoption could happen in few years (Randall 2016) pushed by:

- Governments (international, regional, national or local) which would like to lower CO_2 emissions,
- Rising awareness of customers for a silent and emission free individual mobility,

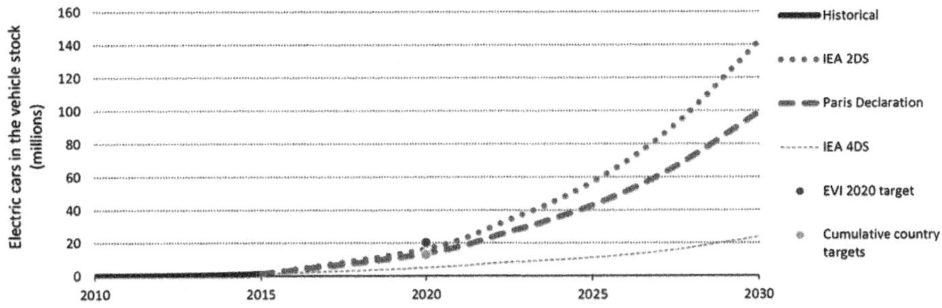

Fig. 3.5 Scenarios for the diffusion of EVs until 2030. *Source* OECD/IEA (ed.): Global EV Outlook 2016, Paris 2016, p. 21

- New competitors like Tesla which are valuated today to 26 bn € (compared with 8 bn € of Fiat Chrysler Automobiles e.g. or 23 bn € of Renault) and
- Technological progress which lowers costs and improves range performance of batteries.

It is reasonable to think that in the near future EV will be faster, safer, cheaper and more convenient than ICE cars (Randall 2016). Most of the OEM's and battery producer are working on it even though it is clear that it is a bet on the future. But like showed on the valuation of Tesla the market believes on it.

The Paris Declaration on Electro-Mobility and Climate Change and Call to Action, announced at Paris during the COP21, a global deployment of 100 million electric cars by 2030 (OECD/IEA 2016). The diffusion of EV's should thus looks like in Fig. 3.5. Besides this scenario the International Energy Agency (IEA) developed also a 2DS EV deployment scenario, limiting global warming to 2° and a 4DS scenario, accepting 4° global warming by 2100. Following the recommendation of the Paris Declaration, some countries think about forbidding sales of new ICE vehicles: Norway and the Netherlands from 2025 and India from 2030 (Bottet 2016).

In Sect. 3.3, we were focussing on product innovation. Section 3.4 is dedicated to the question how innovation of the service dominant logic scenario could leverage value propositions.

3.4 Innovation in Mobility Services and Value Added Services to Meet the Challenge: Increasing the Productive Use of Assets and Customer Centric Mobility

The "value of time" (Fraunhofer IAO and Horvath & Partners 2016) is probably the easiest way to understand the fast developments, the raise of innovation and the value proposition in mobility services. Privately owned cars are currently used only

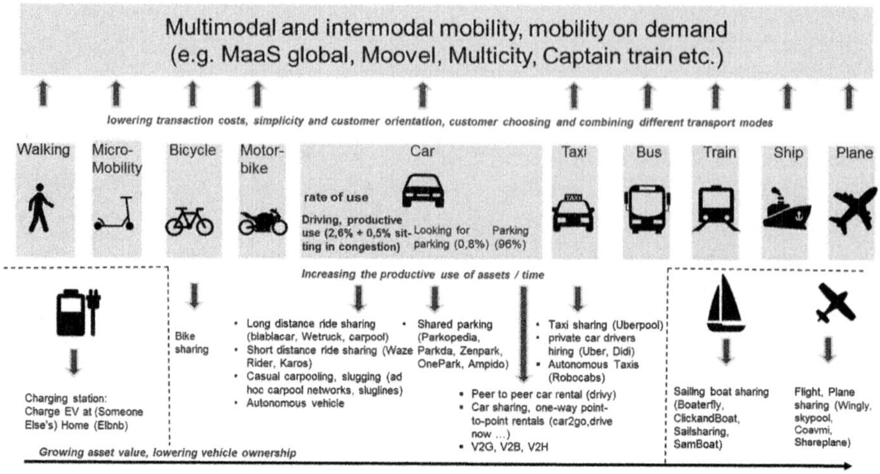

Fig. 3.6 Multimodal Mobility: customer centric, cost and resource efficient mobility services. *Source* Own research

2.6 % of the time for driving; 0.8 % are spent to look for a parking lot; 0.5 % for sitting in congestion; the rest of the time, approximately 96 %, the typical car is parked. Therefore, cars are remarkably underutilised and inefficient (Heck and Rogers 2014; Kalanick 2016).

Digitalisation changed the game through lowering transaction costs (see Sect. 3.2.3), enabling new simple contacts between customer and mobility providers and making the sharing of valuable assets easier: cars and taxis, parking places, bicycles, ships, planes etc. (see Fig. 3.6) can be shared and thus improve the time of their use, minimise the use of energy, materials, place etc. and lower the internal and external costs.

Ride sharing can help improve the efficiency of the driving time by sharing empty seats in a car with passengers through a service platform. Currently there are long distance ride sharing service platforms and others for short distance. "BlaBlaCar" is a long distance ride sharing service with currently about 25 million users.[9] The costs of the ride are shared between driver and passenger and sometimes the platform. The number of users is important in the net economy because of the direct network effects. Network effects mean an enlargement of the network has a direct positive impact for all users (Clement and Schreiber 2013). In the case of carpooling e.g., the probability to get a ride is getting higher with each additional user. Dynamic ride sharing means services can arrange shared rides on a very short notice. Besides these direct network effects, there are also indirect effects (Clement and Schreiber 2013) like user ratings or reputation which lead to an increased quality with each additional user. Casual carpooling systems are different: the aim is

[9]In Europe, Mexico and India, data from BlaBlaCar (2015).

to access to the "3+ high occupancy line" and avoid congestions. This system is existing in the US e.g. in San Francisco, Washington or Houston and focuses on morning pick-up points. Riders queue at these points, as if they would wait for a taxi. Drivers pick up the riders to have more than three passengers in the car. Benefits are the saving of energy, money and time for the users and the assumed positive impact on traffic.[10] Other short distance platforms try to analyse permanently daily habits (time, journeys, etc.) of the user and to mix them with other users in order to offer a driver or a passenger for their journeys to the users. These business models can also offer specific solutions for companies which would like to diminish parking lots and CO_2 emissions for daily journeys to work of the employees. For the user of short car sharing, the cost of travel to work could be diminished in average from 5000 € p.a. to 1500 € (Binois 2016).

Car parks can be seen as a mobility service as well and can be optimised in a similar way as driving. 9.2 % of drivers are ready to renounce to car ownership because of the time loss for searching a parking lot (Millard-Ball et al. 2005). According to Roland Berger, 30 % of the time of city drivers is spent to find a parking lot (Roland Berger 2014). Huge markets will arise to avoid this waste of time and a lot of start-ups already exist in this field. Examples of start-ups are: Parkopedia, Parkda, Zenpark, OnePark, Ampido, sharedparking etc. Via these platforms, owners of parking areas can offer drivers in cities e.g. a parking place at the driver' destination. A revenue of 2 bn € in this field is expected until 2020 by Roland Berger (Tobias 2014).

Sharing vehicles is a further opportunity to use the asset better and save money for customers with small capital lock-up, use less resources (natural resources, surface, time etc.) and decrease pollution impact (Fournier et al. 2015). This is seen as a business opportunity for car manufacturers like BMW and Daimler with DriveNow and car2go. Car2Go from Daimler and Flinkster from Deutsche Bahn, a German train company, have furthermore merged the offer of both companies. Customers will have access to 7000 cars in Germany. Car sharing is one of the first business models of "collaborative consumption" and growing very fast since the seventies. The car sharing market in Germany e.g. started in 2001 from less than 50,000 users and has been growing to more than 1.2 million in 2016. Internet and the introduction of free-floating car sharing are key factors in this fast development. Station-based car sharing services require members to pick up and return vehicles from/to a particular place. With one-way or free-floating services, the car can be found with an app and has not to be brought back to a lot which may be more convenient. More than 830,000 people are using free-floating services, 25 % more than in 2015 (Bundesverband Carsharing 2016). Pear to Pear (P2P) car sharing could experience a fast growth in future, too. P2P car sharing is an online private car rental service, which allows people to rent their private cars to others. The French start-up Drivy just acquired Buzzcar in France and autonetzer.de in Germany. In France in 2015 the service had 500,000 users and 27,000 vehicles to

---

[10]For more details about casual Carpooling locations and characteristics see Minett (2011).

rent and in Germany 100,000 users and 11,000 car owners. Opel (carunity) and Ford (getaround) are now entering this market as well. Roland Berger announced for 2020 a revenue of €3.7–5.6 bn € for car sharing (Roland Berger 2014).

The better use of vehicles assets could also be improved through combining carsharing and autonomous vehicles. Two scenarios are possible: on one side self-driving cars can be shared simultaneously by several passengers (Taxibots) while in another scenario single passengers are pick-up and drop-off sequentially (Autovot) (Burns et al. 2013). A large-scale uptake of a shared and self-driving fleet of vehicles could remove in the taxibots scenario 90 % of the cars in European mid-sized city, keeping a nearly same level of mobility. In the autovot scenario, nearly 80 % of the cars of the city could be removed. Significant public and private space could be free up. This would impact furthermore congestion depending on system configuration and especially on the provided public transport.

Other scenarios were calculated for autonomous vehicles in New York: the replacement of New York's 13,000 yellow cabs with 9000 self-driving ones is capable of supplying better mobility to radically lower costs per mile due to fewer taxis, less empty miles and reduced labour costs of the driver: costs could be lowered by 87 % and waiting time by more than 80 % to less than 1 min (Burns et al. 2013). Robocabs (autonomous battery electric taxis) have according to another study a disruptive potential also to reduce greenhouse gas (GHG) emissions by 87–94 % (in comparison with conventionally driven vehicles in 2014) by 2030 (Greeblatt and Saxena 2015). Google, Apple, Uber, Daimler, Navya etc. are working on autonomous vehicles. Google e.g. would like to prove through autonomous vehicles that artificial intelligence can be superior to human brain and can cause less accidents. Apple would like to overcome the current complexity of cars which might be too complex for average customers. According to AT Kearney, autonomous vehicles will be available on industrial level by 2030 and be implemented first in California, then in China where customer are more likely to buy equipment for autonomous driving then in Germany (Cornubert 2016). A market of 95 bn € is expected for driver assistance systems and additionally 79 bn € for software features (Lagarde 2015).

Uber is a typical example of providing on-demand mobile services for urban transportation on the way probably to driverless transport (Cornubert 2016). Uber develops, markets and operates applications which allows consumers to get a car ride from drivers who use their own cars or other vehicles like helicopters (Samuel 2015). They are today in competition with taxis but innovate also in alternative mobility services autonomous vehicles (see above). The company was founded in 2011 in San Francisco and is now active on all continents. Uber operates in more than 300 cities and has more than 1 million drivers: 26,000 in New York, 15,000 in London, and 10,000 in Paris (Samana 2015). The market capitalisation raised in 5 years to $ 60 bn (€55 bn) (Kresge 2016). In comparison Nissan was evaluated at the same time (June 2016) at 40 bn €, Ford at 45, GM at 40 and Daimler at 63 bn € (fianzen.net 2016). Didi, a Chinese competitor to Uber, has 300 million users and provide 14 million journey per day with 15 million drivers (Rousseau and Ruello 2016).

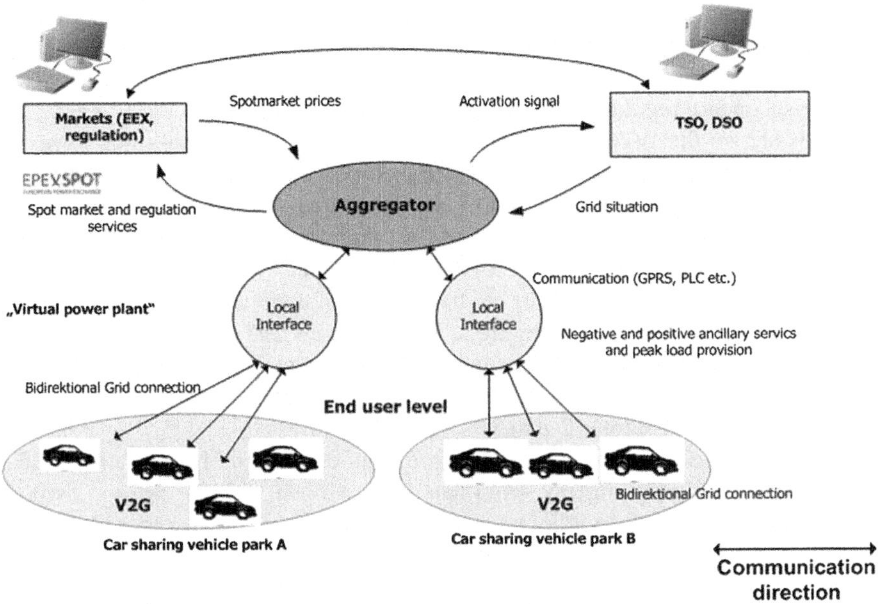

Fig. 3.7 Schematic V2G, Aggregator, Carsharing and energy market interaction. *Source* Fournier et al. (2014, p. 72)

On-demand mobile transportation services can also be used to provide other vehicles like Sailing boat (see sites of: Boaterfly, ClickandBoat, Sailsharing, SamBoat), Plane (see e.g. sites of Wingly, skypool, Coavmi, Shareplane) or even helicopter (Grasland 2016). The motivation is very similar: improving the overall utilisation rate of assets. The rate of utilisation of a helicopter e.g. is according to Airbus just 400 h in a year (Grasland 2016).

To raise the efficiency of use of vehicles, cars can be shared also with the grid.[11] The EV is then connected to the grid and delivers mobility and additionally energy services. V2G (vehicle-to-grid) means that vehicles are used as flexible energy storages, which can compensate peak loads and off-peak periods or which can be used for frequency regulation of the power grid. With this intelligent load and power management, or two-way delivery of power, it is possible to use, for example, the surplus from renewables stored in electric vehicles to substitute peak–load electricity generation normally provided by non-renewable power plants. V2G could then support the integration of renewable energies within e.g. the German Energiewende (Energy transition) into the grid and help lowering CO_2 emissions and enhance grid stability. Furthermore, the aggregated vehicles (e.g. car sharing fleets), which can be thought of as a virtual power plant, can be utilized to generate revenue through administration of frequency and voltage (load-frequency control

[11]For more details see also in the following: Fournier et al. (2010, 2014).

and performance) (see Fig. 3.7). Carsharing EV-fleets could generate additional revenues in participating to these markets when EVs are parked and not booked by customers. Results show, that potential profit only appears when participating to the negative secondary regulation market. With a median value of 173 € per vehicle per year, it is shown that V2G could make economically sense but does not solve the profitability problematic of carsharing with electric vehicles.

Bikesharing becomes more and more popular in many European cities and in the USA. Cycling is generally perceived as a healthier, environmentally friendly, cheaper and often quicker way to move in a town than by public transport or car. The city of Copenhagen is targeting for the number of commuters by bike to reach a 50 % share of inner-city traffic by 2025 (Copenhagen 2011). Using new internet technologies will strengthen the existing offers of railways, municipalities, outdoor advertising companies etc. Roland Berger estimate the market to 3.6–5.3 bn for bike sharing (Roland Berger 2014).

Several services can also be provided from single source mobility platforms like moovel, a subsidiary company from Daimler. Moovel brings car2go, mytaxi, Flinkster, train services, public transport, taxi services, car-shares and bicycle options into one simple application. The idea is to become the "Amazon of mobility". The user registers once and has access to the full range of services for one invoice. The platform has currently about 1 Million users. The aim of the service provider is to become a platform for mobility like Google is for search or Amazon for online retail sales. Moovel bought in 2014 the mobility platform Ridescout in USA. To provide mobility platforms to the corporate mobility segment is in focus as well (Heching 2014).

Improving the value of time and the use of assets, economising costs, extending the mobility means, easing the access to mobility, meeting people, saving space in cities and emphasising environmental aspects offer to customers thus an interesting new value proposition.

3.5 Innovation in Developing Countries to Meet the Challenge: Jugaad Innovation and Reverse Innovation in a Frugal Economy

As pointed out in the introduction of this chapter, closing the gap between developed and developing countries is a huge driver of growth in general and for the mobility sector in particular.

To close the gap, emerging economies have to deal with less resources. Furthermore, buyers in poor countries demand a completely different price-performance ratio (value proposition) for goods and services (Govindarajan 2011). In order to overcome this lack and meet the needs of developing countries, a method called jugaad innovation was developed and experienced in India to face harsh constraints by improvising an effective solution using limited resources

(Prabhu and Jain 2015; Radjou et al. 2012; Radjou and Prabhu 2015). Jugaad describes the frugal, flexible, and inclusive approach to innovation and entrepreneurship. It is adapted to emerging countries but could also be exported to developed countries. In this case it is called reverse innovation: this means when design is local and the product made in emerging countries and for emerging countries and in a next step for the world. The strategy differs from current globalisation strategies where companies of developed countries provide products for developed and emerging markets and from glocalisation strategies where companies develop products in developed countries and fabricate them in emerging markets for the emerging markets.

An example of jugaad innovation transposed to the automobile industry can be given with the Kwid in India. The Kwid is a low cost representative and city suitable SUV (Sport Utility Vehicle) designed by Indian Engineers with French and Japanese know-how. The chosen strategy from Renault Nissan in this case was a so called design to cost strategy in opposition to the carry-over strategy used together with Mahindra in 2007 or ultra-low cost strategy developed with Bajaj in 2008 (Ducamp 2015a, b). The goal of the carry-over strategy was to use already amortised parts from the group in a similar way to Dacia with Renault. The aim of the ultra-low-cost strategy was to develop something like a modern rickshaw. Design to cost by contrast sets limits to costs and weight. Hence, with the help of a local supplier, investments were confined to 0.42 bn €, 50 % less than the usual investments in R&D for a new vehicle. 97 % of the value was created local and the sale price of the vehicle amounts to 4200–5500 € (Ducamp 2015a, b). With a weight of less than 800 kg this family car saved parts and materials. This is a virtuous circle: a part which is not used is the cheapest part in a vehicle; at the same time the vehicle became the most fuel-efficient petrol vehicle in India with 25.17 km/l (Sunny 2015) (kilometre per litre, this corresponds to 3.97 l/100 km). Frugal innovation thus helps to save costs and weight. The Kwid which was developed in India will probably be produced in Brazil and other developing countries in the future and in a next step be exported to emerging and mature markets (reverse innovation).

In China, in a similar way, low speed electric vehicles (LSEV) are developed. The market is growing very fast with the support of LSEV producers, the local government and the customers. Low speed electric vehicles have a simple structure and are equipped with low cost low energy batteries. They meet the mobility needs of rural customers who are looking for a very cheap price, low maintenance costs and multipurpose use adapted to rural road conditions (Shen and Feng 2016). It seems that LSEV have a bigger market share in China than "new energy vehicles" which are promoted from the central government of China.

To conclude, emerging countries will catch up to the living standard of developed countries. This certainly questions the availability of energy and resources, the problem of pollution or the limits of urban systems. However, through jugaad innovation and reverse innovation "emerging countries are pioneering the art of frugal engineering" and "provide practical tips on how Western companies can likewise do more with less" (Ghosn 2016) to tackle this situation.

3.6 Consequences for Value Proposition and Value Chain

Innovation in products can help improve the efficiency of using energy of mobility and reduce CO_2 emissions on a well-to-wheel basis if renewables are used. This is a technical approach where the former ICE powertrain will be successively replaced by batteries. The value for the customer will be in meeting the new legal constraints against global warming as well as in travelling and commuting cheaper in the long run. This will impact the value chain of raw materials, parts and components, the vehicle design, development, production and sales and challenge traditional OEM's which shaped until now the value chain through control of ICE technologies. New players like Tesla will arise and focus more on software and content in tomorrow's value of the car. The hardware will decrease from 90 % today to 40 %. Software will be 40 % and content 20 % of the value creation (Cornubert 2016). The infrastructure will have to be further adapted in order to provide electricity to fuel the vehicles.

Moreover, it is assumed that the importance of the recycling of parts of the vehicle in general and the drive train in particular will increase. Thus the re-use or recycling is a central contribution of sustainable and successful businesses due to the scarcity of numerous raw materials. For example, the 4R Energy Corporation of Nissan and Sumitomo can be mentioned here, as it pursues the re-use, recycling, processing and resale of lithium ion batteries trying to keep track and control of the resources used.

Customer could gain in purchasing power, meet people and protect the environment with innovation in mobility services: Hoibian and Daudey (2014) Purchasing power is reinforced through sharing expensive assets and saving time[12] in traffic jam or looking for a parking slot; contacts with people outside the family circle are enabled and strengthened; sharing assets can finally reduce the footprint of mobility through a better utilization of resources and energy, decongest cities, free up urban space and lower pollution. With autonomous vehicles, travelling and commuting could also be safer if artificial intelligence becomes superior to human brain. The net economy can furthermore help drivers or owner of vehicles to participate to mobility services (see Fig. 3.8) and be part of the value created. The users of these peer-to-peer services become prosumer (Ritzer 2010): producer (e.g. driver) and consumer (e.g. passenger), as the case may be.

Bundled services could aid to match people headed in the same direction at the same time and integrate different transportation options into a seamless experience (Kalanick 2016). The customer would be able to tailor the most efficient combination of ride-sharing, peer-to-peer car rental, bicycle, public transport, train, plane etc. depending on his preferences and needs. The value proposition for customer seems thus to be considerable through single or bundled mobility services.

[12]An evaluation of monetarisation of saving through autonomous vehicles can be found in Fraunhofer IAO and Horváth & Partners (2016).

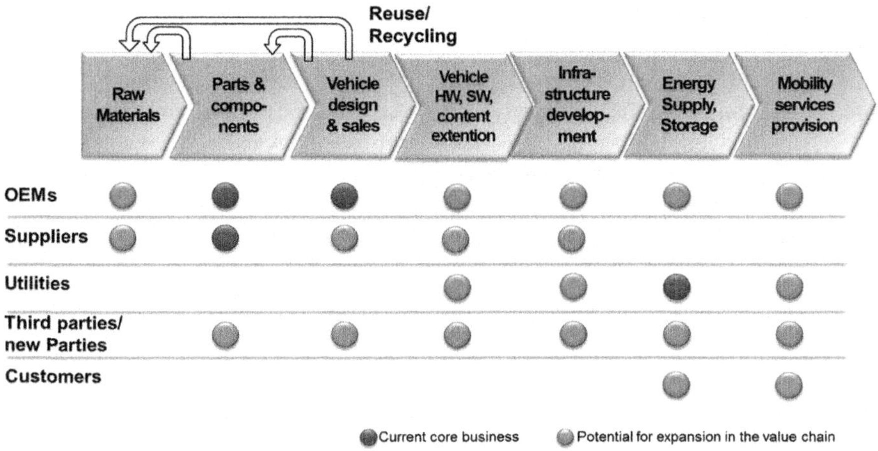

Fig. 3.8 New value proposition and value chain. *Source* Fournier et al. (2012)

Frugal Innovation will finally save energy and resources through incremental innovation and meet the needs from developing countries. Later it can be expected that developed countries will be attracted by reverse innovation as well to make individual mobility more affordable also in developed countries.

Innovation in product and services as well as jugaad innovation have the potential to be disruptive and to arouse technical and social progress but will also impact the offers and the value chains (see Fig. 3.8) to better meet the needs of the customer. Due to the expected developments in electrifying the drive train described above, access to raw materials will gain a crucial role in global supply. E.g. the predominant use of lithium-ion batteries as energy source for the electric motors leads to greater demand for lithium or lithium carbonate as well as for rare earth elements, e.g. lanthanum for the construction of accumulators or neodymium for the production of efficient electric engines. Additionally, further new components, such as power electronics (e.g. inverters) or auxiliary systems, in electrical automobiles become necessary. Moreover, the provision of energy and of the increasingly demanded charging infrastructure for electric vehicles, changes the link of the value chain with the energy supply and the energy suppliers fundamentally.

New players (see Fig. 3.8) and shapers in the value chain (using economies of scale, economies of scope, network and lock-in effects but without profits) will arise with mobility services and challenge traditional (profit oriented) car manufacturer focused on parts, vehicle design and sales (Fournier and Donada 2016). Incumbents of the automotive industry could be threaten by these new players which offer single or bundled mobility services, snap up customer and capture more value from them. In fact, only the companies, which are in touch with customer, can understand their needs and accordingly offer the best solution for them (Unkonwn 2015; Heinemann 2015). OEM's could then be the losers if the service dominant logic scenario is confirmed and if they do not adapt their business models. Examples of other industries challenged by internet companies (music, newspaper, hotels, etc.) show that a fast adaptation and new value creating fields must be found (Dawson et al. 2016). Some of the automotive OEM companies have already reacted and have acknowledged that digitalisation revolutionizes the work of automakers across the entire value chain. More and more cooperations have thus been agreed between OEMs and IT companies e.g. Toyota with Uber, General Motors with Lyft, Volkswagen and Gett etc. (Unkonwn 2016). For the CEO of Daimler this revolution offers great opportunities. Therefore, the question for him is: "how do we manage to be driver and not driven" (Schirg 2015).

3.7 Conclusion

Global warming, the scarcity of and dependency on fossil fuels and other external effects—as well as social changes due to increasing urbanization and changes in values—show the limits of the current mobility paradigm and can be identified as main drivers of changes in the mobility sector. Furthermore, on a more general level the current linear economy reached its limits as well and needs a transformation towards a circular economy which will impact the mobility sector additionally. These trends trigger regulatory interventions on the global, regional, national and local level and therefore enforce the change of the framework conditions, leading to a new socio-economic mobility paradigm which causes changes in technology, value propositions and business models.

As a result, innovation in powertrains will be promoted by all the economic actors. It seems that in the long run, electric vehicles represent the most promising alternative technology because they are the most energy efficient and have the lowest CO_2 emissions on a well-to-wheel basis. This product-oriented approach is necessary but will be completed by a service oriented holistic approach, which provides mobility services triggered by digitalisation. The induced innovations will diffuse rapidly because it meets the customer needs more extensively, and will make mobility more affordable and more environmental friendly. Jugaad innovation could finally meet the needs of developing countries but also subsequently of developed countries through reverse innovation. This incremental innovation will reduce the gap between rich and poor countries and, using less resources and energy, help to pave the way to a sustainable frugal economy and to a sustainable mobility.

The value chain and value propositions of mobility solutions will therefore be deeply impacted in the future: new raw materials, components, vehicles and services will arise; new players will reshape the value chain and challenge traditional OEM's with new product and services; through customer centric approach the new shapers can seize customer's interests and contact, and thus capture more value; customer will even be part of the value chain and become prosumer, being either consumer or producer as the case may be.

Innovation can thus give an answer to the current systemic challenges and externalities our societies are facing. The new mobility paradigm is a part of this transformation towards sustainability which will transform value proposition and value chain. A new wave of innovation with technologies reducing, instead of enhancing, resource consumption and environmental degradation could raise (Weizsäcker et al. 2016) but it is unclear whether these transformations will be adopted and whether they will trigger a new wave of growth or a new Kondratiev cycle (so called new techno-economic paradigm) (Freeman and Perez 1988) like it could be expected in theory: the near zero marginal costs for digital goods and services as well as the development of prosumer could make it difficult to measure evolution in wealth; furthermore a frugal economy which can be lined with degrowth changes the perception of wealth and raises the central question what is the right level of consumption in order to live a full life (Latouche 2010; Levallois 2010; Bafina and Ravishankar 2014) In fact, sustainable mobility and progress in environment, economics or social matters are feasible but will probably be difficult to measure and to prove.

Acknowledgments I would like to thank António Moniz, Christina Merz, Roukaya Issaoui and Iuliana Ilie for their dedicated support in preparing this manuscript and my colleagues and students of the School of Engineering and the Business School for our fruitful discussions.

References

Avere-France, 2016. *Avere France.* [Online] Available at: http://www.avere-france.org/Site/Article/?article_id=6536

Bafina, N. & Ravishankar, S., 2014. The need for change in the Approach towards global change and environment protection plans to secure sustainable development. *OIDA International Journal of Sustainable Development*, 22 January, 06(07), p. 10.

Binois, E., 2016. Karos ou la démocratisation du covoiturage courte distance. *autoactu.com.*

BlaBlaCar, 2015. keynote at google I/O 2015; Moscone Center West, San Francisco, CA. May 28–29.

Bottet, G., 2016. *Leblogauto.com.* [Online]. Available at: http://www.leblogauto.com/2016/06/interdiction-des-moteurs-essence-et-diesel-en-norvege-en-2025.html. [Accessed 06 June 2016]

BP, 2015. *BP Statistical Review of World Energy*, London: BP.

BP, 2016. *BP Energy Outlook to 2035*, London: s.n.

Bundesministerium für Umwelt, Naturschutz, Bau und Reaktorsicherheit,BMUB, 2009. *Die EU-Verordnung zur Verminderung der CO2 - Emissionen*, s.l.: s.n.

Bundesverband Carsharing, 2016. *CarSharing-Jahresbilanz 2015*, Berlin: s.n.

Burns, L., Jordan, W. & Scarborough, B., 2013. *Transforming personal mobility*, Broadway NY: Columbia University.

Chamon, M., Mauro, P. & Okawa, Y., 2008. Mass car ownership in the emerging market giants. *Economic Policy*, 1 April, 23(54), pp. 271-272.

Christensen, C., 1997. *The innovators dilemma: When new technologies cause great firms to fail.* Boston: Harvard Business Press.

Clement, R. & Schreiber, D., 2013. *Internet-Ökonomie Grundlagen und Fallbeispiele der vernetzten Wirtschaft.* 2 ed. s.l.: Springer-Gabler.

Concawe; EUCAR; European Commission JRC, 2007. *well to wheel analysis of future automotive fuels and powertrains in the European context*, s.l.: s.n.

Copenhagen, C. o., 2011. *Good, Better, Best- the City of Copenhagen's bicycle strategy 2011-2025*, Copenhagen: the City of Copenhagen Technical and Environmental Administration.

Cornubert, R., 2016. *série de viédo Voiture autonome*, s.l.: autoactu.com.

Dalan, V. M. & Dol, N., 2010. Junge Deutsche verlieren Lust am teuren Auto. *Die Welt*, March.

Dawson, A., Hirt, M. & Scanlan, J., 2016. *The economic essentials of digital strategy*, s.l.: Mckinsey.

Décret n° 2016-703, 2016. *décret relatif à l'utilisation de pièces de rechange automobiles issues de l'économie circulaire.* [Online] Available at: https://www.legifrance.gouv.fr/eli/decret/2016/5/30/DEVP1602645D/jo/texte

Directive 2000/53/EC of the European Parliament and of the Council, 2000. *end of life vehicles*. s. l.: s.n.

Ducamp, P., 2015a. La génèse d'un véhicule mondial. *Usine nouvelle*, 27 May.

Ducamp, P., 2015b. Quid de la Renault Kwid. *Usine Nouvelle*, 20 May.

Dupond-Calbo, J., 2015. Les raisons du divorce entre les jeunes et l'automobile. *Les Echos.*

Durand, D., 1993. *La systèmique.* 12 ed. Paris: Presse Universitaire de France.

EIA, 2016. *International Energy Outlook 2016*, s.l.: EIA.

Ellen Macarthur Foundation, 2013. *Towards the Circular Economy-Opportunities for the consumer goods sector*, s.l.: s.n.

Ellen Macarthur Foundation, 2014. *Towards the circular Economy-Accelerating the scale-up across global supply chains*, Geneva: World Economic Forum.

Ellen MacArthur Foundation, 2015. *Towards a circular Economy: Business Rationale for an accelerated Transition*, s.l.: Ellen MacArthur Foundation.

Energy Watch Group, 2007. *Crude Oil-The supply Outlook*, Ottobrunn: Energy Watch Group.

fianzen.net, 2016. www.finanzen.net. [Online] Available at: http://www.finanzen.net/aktien/Nissan-Aktie

Fournier, F. & Donada, C., 2016. Future Business Models and shapers for the Automotive Mobility?. In: *Nationale und Internationale Trends in der Mobilität-technische und betriebwirtschaftliche Aspekte.* Wiesbaden: Gabler Verlag, pp. 27-39.

Fournier, G., 2009. How to cope with the distance in the future? the impact of Globalisation and Ecological requirements the Destiny of the automotive industry and its suppliers. *Economic and Business Review*, February, 11(1), pp. 75-100.

Fournier, G., Baumann, M. & Seign, R., 2010. Integration von Elektrofahrzeugen in ein Netz mit hohen Anteil an erneuerbaren Energien.Mögliche ökonomische und ökologische Auswirkungen. *Zeitschrift für die gesamte Wertschöpfungskette Automobilwirtschaft*, pp. 30-43.

Fournier, G. et al., 2014. Carsharing with Electric Vehicles and Vehicle-to-Grid: a future business model?. Dans: H. e. a. Proff, éd. *Radikale Innovationen in der Mobilität - technische und betriebswirtschaftliche Aspekte.* Wiesbaden: s.n., pp. 63-80.

Fournier, G. et al., 2012. The new mobility paradigm: Transformation of value chain and business models. *Entreprise and Work innovation studies*, December, 8(8), pp. 9-40.

Fournier, G., Seign, R., Goehlich, V. & Bogenberger, K., 2015. Car-sharing with electric vehicles: A contribution to sustainable mobility?. In: *Interdisziplinäre Managementforschung XI.* Opatija: s.n.

Fournier, G. & Stinson, M., 2011. The future thinks electric. Developing at electric mobility value chain as a foundation for a new energy paradigm. *Briefly,R./Turkalj, Z.(Hrsg.): Interdisciplinary management Research VII;*, pp. 868,875,876,883.

Fournier, G., Würzer, D., Seign, R. & Hinderer, H., 2012. Der Weg zur emissionsfreien Moblität-Strategien der Hersteller im internationalen Vergleich. In: *Zukünftige Entwicklungen*

in der Mobilität- Betriebswirtschaftliche und technische Aspekte. s.l.: Springer Gabler, pp. 405-423.
Fraunhofer IAO and Horvath & Partners, 2016. "*the value of time*", *Nutzerbezogene Service-Potenziale durch autonomes Fahren*, Stuttgart: s.n.
Freeman, C. & Perez, C., 1988. Structural Crises of Adjustment, business Cycles and Investment behavour. In: *Technical change and economic theory*. London: France Pinter, pp. 38-61.
Freyssenet, M., 2009. The second automobile revolution: promises and uncertainties, in: Freyssenet, M., (ed.), The Second Automobile Revolution. Trajectories of the World Carmakers in the 21st century. Dans: Basingstoke and New York: Palgrave Macmillan, pp. 443-454.
Ghosn, C., 2016. *Jugaad in action*. [Online] Available at: http://jugaadinnovation.com/supporters/
Globim, T. a. E., 2016. *The basis for Biofuel policy post 2020*, s.l.: s.n.
Govindarajan, V., 2011. Five Phases Toward "reverse Innovation". *Wall Street Journal*, 04 December.
Grasland, E., 2016. Paul Eremenko: "Notre mission, boulverser Airbus et l'aéroanutique tout entière". *Les Echos*, 10 May.
Greeblatt, J. & Saxena, S., 2015. Autonomous taxis could greatly reduce greenhouse-gas emissions of US light-duty vehicles. *Nature Climate change*, Volume 5, pp. 860-863.
Heching, M., 2014. Daimler-Tochter Moovel: "Wir wollen das Amazon der Mobilität werden). *Manager Mazin*, 26 September.
Heck, S. & Rogers, M., 2014. Are you ready for the revolution?. *McKinsey Quarterly*.
Heinemann, G., 2015. Digital zestört alte Kundenbeziehungen. *Frankfurter Allegemeine Zeitung*, 26 October.Volume 248.
Hoibian, S. & Daudey, E., 2014. La société collaborative: mythe et realité. *Cahiers de recherche du Crédoc*, Issue 313.
IEA, 2010. *World Energy Outlook 2010 Executive Summary*, Paris: International Energy Agency (IEA).
IEA, 2011. *Technology roadmap Biofuels for transport*, Paris: s.n.
Intergovernmental Panel on Climate Change, 2007. *Synthesis report*, Valencia: Intergovernmental Panel on Climate Change.
Intergovernmental Panel on Climate Change, 2014. *Summary for Policymakers*, Cambridge/New York: Intergovernmental Panel on Climate Change.
International Transport Forum, 2011. *Tranasport outlook 2011: meeting the needs of 9 Billion People*, Paris: OECD/ITF.
Kalanick, T., 2016. Uber CEO Travis Kalanick's Gridlock Solution? Carpools For All, the founder and CEO of the ride sharing juggrnaut on why carpooling isn't just for school children. *The Wall Street Journal*, 6 June.
Ki-moon, B., 2011. *Secretary-General's remarks to the World Economic Forum Session on Redefining Sustainable Development*. Dovos Switzerland, United Nations.
Kresge, N., 2016. Uber CEO wants to partner with automakers, not build cars. *Bloomberg*, 08 June.
Lagarde, F., 2015. *Les constructeurs bien placés pour gagner la course de la voiture autonome*. s. l.: autoactu.com.
Latouche, S., 2010. Editorial: Degrowth. *Journal of Cleaner Production*, pp. 519-522.
Levallois, C., 2010. Can de-growth be sonsidered a policy option? a historical note on Nicholas Georgescu-Roegen and the Club of Rome. *Ecological economics*, pp. 2271-2278.
LOI n° 2015-992, 2015. *loi relative à la transmission énergitique pour la croissance verte*. [Online] Available at: https://www.legifrance.gouv.fr/eli/loi/2015/8/17/DEVX1413992L/jo/texte
Maibach, M. et al., 2008. *Handbook on estimation of external costs in the transport sector: Internalisation Measures and Policies for All external Cost of Transport (IMPACT) Version1.1*, Delft: European Commission.
Malthus, T. R., 1798. An essay on the principle of population as it affects the future improvement of society, London.

Millard-Ball, A. et al., 2005. *Car-Sharing Where and how it succeeds: TCRP report 108*, Washington,D.C: Transportation research board.
Minett, P. &. P. J., 2011. Estimating the energy consumption impact of casual carpooling. Energies, 4(1), 126-139, p. 129.
Nations, United, 2010. *World Urbanization Prospects*, New York: United Nation.
Nykvist, B. & Nilsson, M., 2015. Rapidly falling costs of battery packs for electric vehicles. *Nature Climate Change*, Volume 5, pp. 329-332.
Nysveen, P. M., 2015. 2015 will be extraordinarily tough for oil companies, Energy research and analysis, Rystad Energy, January 12, 2015.
OECD, 2014. *The Cost of Air Pollution: Health Impacts of Road Transport*, s.l.: OECD Publiching.
OECD/IEA, 2016. *Global EV Outlook 2016,* Paris: s.n.
Optiresource-Software, n.d. *Daimler Optiresource.* [Online] Available at: http://www2.daimler.com/sustainability/optiresource/pages/pagLanguageSelect.html
Parin, M., 2016. *Breezcar le marché des hypbrides et électriques.* [Online] Available at: http://www.breezcar.com/actualites/article/tesla-model-3-electrique-135000-precommandes-monde [Accessed 02 April 2016].
Perez, C., 1983. Structural change and the assimilation of new technologies in the economic and social system. *Future Nr. 4, Bd. 15*, October, pp. 357-375.
Porter, M. & Heppelmann, J., 2014. How Smart Connected Products are transforming competition. *Harvard Busines Review*, Volume 11, pp. 64-88.
Prabhu, J. & Jain, S., 2015. Innovation and entrepreneurship in India: Understanding jgaad. *Asia Pacific Journal of Management*, December, 32(4), pp. 843-868.
Radjou, N. & Prabhu, J., 2012. Mobilizing for growth in emerging markets. *Sloan Management Review*, 53(3), pp. 81-88.
Radjou, N. & Prabhu, J., 2015. *Frugal innovation: how to do more with less.* s.l.: The economist; Public affairs; Profile books.
Radjou, N., Prabhu, J. & Ahuja, S., 2012. *Jugaad innovation: Think frugal, be flexible, generate breakthrough growth.* San Francisco: Jossey Bass.
Randall, R., 2016. *Bloomberg.com.* [Online] Available at: http://www.bloomberg.com/features/2016-ev-oil-crisis/ [Accessed 25 February 2016].
Ricardo, D., 1822. On Protection To Agriculture, 4th edition, London.
Rifkin, J., 2014. *Die Null Grenzkosten Gesellschaft- Das Internet der Dinge, Kollaboratives Gemeingut und der Rückzug des Kapitalismus.* Frankfurt am Main: Compus Verlag.
Ritzer, G. J. N., 2010. Production, Consumption, Prosumption, the nature of capitalism in the age of the digital 'prosumer'. *Journal of consumer culture*, March, 10(01), pp. 13-36.
Roland Berger Strategy Consultants, 2016. Integrated Fuels and Vehicles Roadmap to 2030+. Dans: Munich: s.n.
Roland Berger, 2014. *Shared Mobility: How new businesses are rewriting the rules of the private transportation game*, s.l.: s.n.
Rousseau, Y. & Ruello, A., 2016. l'argent coule à flot en Chine pour Didi dans sa bataille contre Uber. *Les Echos,* 13 June.
Samana, P., 2015. Uber, 5 ans, 1 million de chauffeur, 50 milliards de dollars. *BFM business*, 06 June.
Samuel, H., 2015. Uber offers helicopter taxi service in Cannes to beat competition. *The telegraph*, 13 May.
Schirg, O., 2015. Digigtale Revolution bietet große Chancen. *Hamburger Abendblatt*, 30 July.
Schoettle, B. & Sivak, M., 2014. The reasons for the recent decline in young driver licensing in the United States. *Traffic injury prevention*, 15(1), pp. 6-9.
Schumpeter, J., 1950. *Kapitalismus, Sozialismus, Demokratie.* Bern: s.n.
Schwab, C., 2016. *The Digital Transformation of Industries* [Interview] 2016.
Senat: commission d'enquête, 2015. *le coût économique et financier de la pollution de l'air: rapport remis à Monsieur le Président du Sénat le 8 Juillet 2015, Dépôt publié au Journal Officiel- Edition des Lois et Décrets*, s.l.: Senat.

Shen, Q. & Feng, K., 2016. *The technological tragectories and policy influences in the development of EV in China.* Bonn, IRENA.

Smith, A., 1776. *An Inquiry into the Nature and Causes of the Wealth of Nations, reprinted by the Oxford University Press 1976.* s.l.: s.n.

Spath, F., 2016. *Breezcar le marché des hybrides et électriques.* [Online] Available at: http://www.breezcar.com/actualites/article/ventes-tesla-motors-allemagne-2015-superieures-mercedes [Accessed 20 April 2016].

Stan, C., 2008. *Alternative Antriebe für Automobile; Hybridsysteme, Brennstoffzellen, Alternative Energieträger.* 2, erweiterte Auflage ed. Berlin: Springer.

Sunny, K., 2015. Renault Kwid most fuel-efficient petrol car in India. *Inernational Business Times*, 29 September.

Teichmann, G., 2011. *Elektromobilität- den Anforderungen der Kunden gercht werden*, Nürtigen: PWC.

Tobias, S., 2014. *Shared Parking: Weltweiter Markt wächst bis 2020* [Interview] 2014.

United Nation, 2015. *Framwork Convention on Climate Change: Conference of the Parties 21.* Paris, s.n.

Unknown, 2009. *KBA bestätigt: BMW Group führend bei Verbrauchsreduzierung*, www.Autosieger.de. [Online] Available at: http://www.autosieger.de/article17035.html [Accessed 2016].

Unknown, 2010. *KONGRESS - DIE REDE VON BUNDESUMWELTMINISTER RÖTTGEN*, www.auto-motor-und-sport.de. [Online] Available at: http://www.auto-motor-und-sport.de/news/kongress-die-rede-von-bundesumweltminister-roettgen-co2-grenzwert-von-zehn-bis-35-gramm-ab-2040-1828128.html [Accessed 2016].

Unknown, 2014. *Pièce de réemploi: un avenir radieux,* www.apres-vente-auto.com. [Online] Available at: http://apres-vente-auto.com/pieces/21612-pieces-reemploi-avenir-radieux [Accessed 2016].

Unkonwn, 2015. Ein Plädoyer für die digitale Kontakpflege. *Frankfurter Allgemeine Zeitung*, 24 June.Volume 143.

Unkonwn, 2016. constructeurs automobiles et mobilité numérique passent aux choses sérieuses. *les Echos*, 11 June.

Varian, H., 2000. *Markets for Information Goods, an overview of how markets deal with the peculiar properties of information goods.* s.l., s.n.

Wallentowitz, H. & Freialdenhoven, A., 2011. *Strategien zur Elektrifizierung des Antribsstrangs: Technologien, Märkte und Implikationen.* 2, überarbeitete Auflage ed. Wiesbaden: Springer Gabler.

Weizsäcker, E. et al., 2016. *Factor five: transforming the global economy through 80 % improvement in resources productivity.* s.l.:Earthscan.

Wolrd Business Council for Sustainable Developement, 2004. *Mobility 2030: Meeting the challenges to sustainability*, Geneva: WBCSD.

World Energy Council, 2011. *Global Transport Scenarios 2050*, London: World Energy Council.

Wüchner, E., 2007. Neuartige Antriebe. In: *Vieweg Handbuch Kraftfahrzeugtechnik.* 5, überarbeitete und erweiterte Auflage ed. Wiesbaden: Springer verlag, pp. 111-112.

Wyman, O., 2009. *Oliver Wyman Study "E-Mobility 2025": Power Play with electric-cars.* [En ligne] Available at: http://www.e-mobile.ch/pdf/2009/PM_E-Mobility_2025.pdf

Chapter 4
How Public Policies Can Pave the Way for a New Sustainable Urban Mobility?

Danielle Attias and Sylvie Mira-Bonnardel

Abstract This chapter focuses on the role of governments in the transformation of the automotive industry into the ecosystem of sustainable mobility 2.0 and how this transformation is taking place by putting governments as key players in mobility challenges, endorsed with a new structuring role. We shall show how public policies help to move from the old mobility paradigm based on cars proliferation associated with multiple nuisances to a new paradigm embracing new needs and offers designing mobility in a sustainable environment. The States, the European community as well as local authorities are implementing urban mobility policies, leveraged either by incentives (such as inter-modality and tax reduction) or coercion (such as speed limits, urban tolls), imposing global guidelines aiming at reducing noise or pollution.

Keywords Urban mobility policies · Key mobility players and new challenges · Influence of new regulations · Ecosystem of sustainable mobility 2.0

4.1 Introduction

«*Our future mobility will not be shaped within a unique framework but within a combination of various innovations. This future is being mainly built by private actors but public policies must support and guide their progress*» (Chriqui 2010). This introduction to the analysis of forthcoming new nobilities shows the complex transition towards the ecosystem of smart mobility and the impact of public policy

D. Attias (✉)
Laboratoire Génie Industriel/Industrial Engineering Research Department, CentraleSupélec, Université Paris-Saclay, Grande Voie des Vignes, 92290 Chatenay-Malabry, France
e-mail: danielle.attias@ecp.fr

S. Mira-Bonnardel
Ecole Centrale Lyon, 36 Avenue Guy de Collongue, 69134 Écully, France
e-mail: sylvie.mira-bonnardel@ec-lyon.fr

D. Attias (ed.), *The Automobile Revolution*, DOI 10.1007/978-3-319-45838-0_4

in this transition, putting governments as key players in mobility challenges, endorsed with a new structuring role.

If the XXth century boosted individual carbonized mobility, the XXIth century may open the path for a multimodal, low-carbon and connected mobility, towards a new mobility paradigm (Donada 2013), a paradigm urging to rethink not only technology, energy or socio-economical, ecological and political issues but also industrial strategies facing a transforming ecosystem whose limits are expanding far beyond the frontier of the traditional car industry.

This transformation of the automotive industry into the ecosystem of smart mobility still asks the question of sustainable development since smart mobility won't directly resolve neither urban congestion nor security or energy resources management. These issues need appropriate public policies and new public interventions. Yet the question of the type and level of public intervention is still open, different levers are being designed between incentives and constraints.

Therefore in this chapter, we will explain how governments, as various as local, regional or national levels, implement new regulations, coupling incentives and coercion, in order to shape a sustainable urban mobility. We'll show how the transformation of the automotive industry into the ecosystem of electro mobility is taking place by putting governments as key players in mobility challenges, endorsed with a new structuring role.

The chapter is organized with two parts. The first part presents different regulations implemented by local, regional and national governments, as well as taxation, capital-intensive investments or energy policies with regulatory constraints, to illustrate this new urban paradigm examples of innovative urban policies in the world of megacities in Asia, Europe and the USA will be presented as well.

The second part focuses on the example of France. We explain the role of governments in the shaping of the ecosystem of smart mobility with creation and development of opportunities for all actors of mobility, aiming at developing sustainable urban mobility, largely based on digital technologies and disruptive innovations such as the self-driving connected car which force the legislator to revisit regulations.

4.2 The Challenges for Urban Mobility: To Decarbonize and Decongest the Cities

For many years now, major world cities have been facing mobility-related problems, such as bottlenecks, noise, pollution, greenhouse effects all these problems are related to the huge place occupied by cars yet. Worldwide, 700 million cars are on the roads for 7 billion people; 2015 was a year of record sales with over 80 million vehicles sold worldwide. In 2050, world population should reach 10 billion among which 80 % will live in a few megacities. According to Freyssenet (2015), more than 90 million cars have been produced in 2015; the number should increase by 5 % per year for the next ten years. The car production massively developed during the last XXth century and should still increase for the forthcoming years regarding the huge existing needs worldwide. Yet needs are not equal around the world.

The Chinese car market continues to grow at about 10 % per year and the U.S. and Japanese car markets are doing well. In 2014 France counted about 38.2 million vehicles according to the CCFA, an increase of 0.2 % compared to 2013, although economic growth has declined and sales of new vehicles have decreased. Among world cities the overall trend shows still rising car uses even though we notice differences between developed countries and emerging market.

This increase induces of course negative impacts on public health. In fact, a study of the American consulting company Little (2014) shows how urban politicians have to tackle multiple challenges among which mobility constitute a major issue, either for economic development or public health.

The efforts worldwide to reduce greenhouse gases, fine particles and to decongest cities should change the purchasing behavior of motorists and encourage them to less use their car in the city, preferring shared transportation or the use of cleaner vehicles. Is it to say that cars will disappear from urban centers or should they be renewed in order to better fit the new urban landscape?

For a decade now, public policies have been converging to ease the implementation of a smart mobility bridging at least six dimensions (Navigant Research 2016a):

1. Citizen well-being: quality of live, safety, security, air quality, health improvement
2. Environmental sustainability: emissions reduction, improve land use, energy conservation
3. Smart cities: information services, mobility hubs, charging stations, city landscape
4. Economic viability: economic efficiency, global connectivity
5. Smart energy: energy efficiency, V2G integration, EV charging
6. Smart government: investment, policy and regulation, integration and connectivity, smart cities platforms.

4.2.1 Public Measures: Local Regulations and National Policies

The States, the European community and local authorities are implementing urban mobility policies, leveraged either by incentives (such as inter-modality and taxation) or coercion (such as speed limits, urban tolls), imposing global guidelines aiming at reducing noise or pollution: the CO_2 reduction target set by the European Commission is one such example. To meet these challenges, governments are facing the requirement to completely resize urban transports. They make use of various measures, such as national and local regulations and taxation, investments in infrastructure and equipment.[1]

[1]See Table 4.1 in appendix the different measures giving examples of implementation.

The European Commission adopted in 2008 various regulations based on the *energy-climate balance*. The main objective is to set up a common European energy policy to fight against the climate change. Initially, it promoted a clear objective resumed by the "20-20-20", i.e. the *energy-climate balance* should push the rise of renewable energies up to 20 % of the global energy's consumption, help to reduce by 20 % of the CO_2 emissions, and to increase by 20 % the energy efficiency. These objectives should be achieved by 2020. The *energy-climate balance* gave birth to several legislative texts, in particular the regulation 443/2009 which incites to the development of more ecological vehicles (95 g of CO_2 emissions/km in 2020), as well as the directive 2009/28/CE for the promotion of biofuels (at least 10 % of gmpbam consumption in 2020).

In Germany, the federal government promoted the *Energiewende*: the transition from a nuclear and carbonized energy towards a more reliable and sustainable one. The Legislative support for the *Energiewende* was reinforced in 2010 targeting different achievements by 2050: mainly greenhouse gas reductions of 80–95 % and a renewable energy target of 60 % (Buchan 2012).

Different practices are implemented accordingly to countries regulations and culture (Leurent and Windisch 2013). The Delft report (2011) confirms the heterogeneity between European countries by presenting not less than 18 families of financial and non-financial political instruments; each family counts dozens of actions.

In January 2014, the consulting firm Arthur D Little led an evaluation of maturity and performance for 66 cities around the world. The analysis based on 11 criteria such as the modal mix, the weight of public transport, the number of car per capita, CO_2 emissions, average travel time gave a score to each city ranging from 0 to 100. North America's score, 62 points, is below the average (64.4), barely surpassed by South America at 63.6 points, against 71.4 for Western Europe; the highest scores are obtained by Hong Kong and Amsterdam which lie above 80. Only 15 % of studied cities stand above the 75 % mark.

China, the world's biggest car market, suffers tremendously from urban driving, as for instance Beijing counted 7 million motor vehicles in 2015. To overcome the explosion of the number of vehicles, the city of Shanghai implements monthly auctions to get license plates. Thus the local government can control the number of new registered vehicles.

To solve these problems, governments are facing the need to completely resize urban transports. They make use of various measures, such as national and local regulations and taxation, investments in infrastructure and equipment.

Local governments in cities have also adopted regulations in the alternative control or coercion. The example of Curitiba, a Brazilian town south of Sao Paulo, is particularly interesting. Indeed the town council has since the early 70s implemented an innovative policy on public transport that has been imitated by many developing cities in the newly industrialized countries facing pollution, poverty and huge traffic jams. In 1974, the Curitiba's council took the first initiative to boost public transport, after decades of automotive development and created the first bus lanes.

In 1991, at a time when all cities with more than one million inhabitants chose a very expensive and destructive underground, Curitiba opted for a bus system much cheaper but integrating all the advantages of the underground such as (1) specific bus stops on the main roads, (2) limited waiting time, (3) extra-long and bi-articulated buses (270 passengers) driven by a single employee, (4) access facilities for disabled persons, (5) a large network interconnected with circular routes with 340 buses driving on 1100 km of lines that crisscross the city. Connecting terminals hosts services and shops.

As a result, nearly 70 % of the inhabitants of the urban area and 85 % of city inhabitants use public transport to get around (3.2 million people in the metropolitan area and 1.5 million for the city itself). Curitiba invented the concept of Bus Rapid Transit (BRT) or Bus with High Service Level (BHSL). In June 1996, during the second global summit of mayors and urban planners held in Istanbul, Curitiba was rewarded as the most innovative city in the world. The transport system was also rewarded by the English Building and Social Housing Foundation (EBSHF).

The city has experienced strong growth over the past decade and in order to further limit the use of cars, the city's council decided to develop transport by bicycle by launching, late 2013, a plan to build 300 km of cycle paths to be completed in 2016. Finally, Curitiba's council seeks above all to reconcile economic development and sustainable development by developing intermodal transport.

4.2.2 Prevalent Objective: A Better Quality of Living

Following the example of Curitiba many cities have experienced local policies targeting two main objectives: to decarbonize and to decongest the city. The control of CO_2 emissions is based mainly on traffic regulation with the creation of specific areas such as LEZ (Low Emission Zone) or HOV (High Occupancy Vehicle lanes) and the implementation of tolls or traffic ban.

Yet control and coercion still represent the corner stone of transportation policies in cities. Since 2006, the European norms on air quality have been consolidated, giving birth to the LEZ that largely expanded into Europe. The LEZ are the green solutions designed to improve the quality of life for residents and promote awareness of the inhabitants. They serve as a full-scale laboratory for new infrastructure and organized cooperation between public and private actors, as the development of a LEZ is based on an ecosystem bringing together public and private stakeholders for adequate infrastructure design, including intelligent traffic systems (ITS: Intelligent Transport System) for the management and control, development of dedicated lanes for different modes of transportation.

The city of London chose in 2003 the implementation of a downtown toll: only clean vehicles (less than 100 g CO_2 per kilometer) do not pay the fees. This toll induced a real traffic decrease but bottlenecks remain and the air quality is still not optimum. Stockholm in Sweden chose the same traffic control.

In March 2016, almost 2000 LEZ are organized within European cities in nine countries with an unequal distribution: 85 % of the LEZ can be founded in Germany and Italy. The LEZ showcase technological and eco-friendly solutions to improve life quality. They are used as scalable experiments for the design of new urban infrastructures integrating Intelligent Transport System to pilot and control. Moreover the design of a LEZ leans on a dynamic ecosystem including private and public actors (ADEME 2014).

To go a step further, some city governments organised "*car-free neighborhoods*", giving rise to the development of sustainable neighborhoods, aimed at creating a harmonious cohabitation between the different transport modes and, in fact, to minimize the place given to the car in the public arena, whether in parking as in traffic.[2] As an example, in Amsterdam in the Netherlands, the inhabitants organized a poll to ask for the creation of a sustainable neighborhood without any car: 6000 persons accepted the deal; giving birth to the neighborhood GWL-Terrein.

These sustainable neighborhoods are not design with the fantasy of a city without a car, but are rather designed as invisible car neighborhoods. They are based on the principle that cars should not be seen on public space thanks to original solutions regarding circulation and parking. Car traffic and parking are constrained in favor of pedestrian and bicycle traffic. Parking is disassociated from the constructions with privileged access to public transport in the right-line of the new order of priority given to collective transport.

These different sustainable neighborhoods also represent laboratories for the city of the future. Everything must be indeed thought that any journey within this neighborhood is made by non-motorized means and any way, to join other districts nearby, is feasible through other means than the car and mainly through more efficient and reliable means. Finally, if the most profitable neighborhood is a neighborhood without a car, it is necessary for the people to be ready to change their habits, especially since project deals not only with the creation of infrastructure but also with the lifestyle of the inhabitants. The creation of sustainable neighborhoods involves therefore thinking a new framework of life for the current inhabitants, but also for future generations. It implies also considering the proximity as an important asset for the people that can be easily evaluated by energy savings and CO_2 emission reductions.

Creating a sustainable neighborhood obliges to reconsider the land: changing from a valuable asset to a non-renewable resource that has to be managed at best for the future.

---

[2]See Table 4.2 in appendix the examples of three cities: London, Beijing and Los Angeles.

4.2.3 Public Policies and Electric Vehicle Development

Although, for a century, EV have been emerging technologies (Fréry 2000), it seems that prospective scenario actually show exploding trends for the forthcoming decade (Jullien and Pardi 2013). Navigant research report (2016b) show that BEV (Battery-Electric-Vehicle) and PHEV (Plug-in-Electric-Vehicle) sales will reach 3 million by 2024.

The increase in EVs sales depends on 4 factors:

- Technology developments: better batteries, faster charging, wireless charging
- Air pollution concerns: cities regulations on CO_2 emissions
- Business model innovation: shared mobility leveraging EVs development
- Policies and Regulations: fiscal and financial incentives.

Around the world, financial incentives for the purchase of an electric vehicle are expanding and start to show some results with a real increase in sales, even if the market is still narrow.

The Tokyo's council started early 2000s to help Japanese drivers to by electric or hybrid cars with financial incentives. This policy has proved fruitful since, on the one hand, diesel cars almost disappeared in Japanese cities and, on the other hand, car makers made huge efforts to design less polluting car with electric, hybrid or even hydrogen engines. Meanwhile, diesel vehicles have drastically decreased not only in Tokyo but in the whole archipelago.

The Chinese government has also decided to boost electric cars sales: it forecasts 5 million cars sales with alternative engine (electric or hybrid) in 2020 and big investments on national electric infrastructure have been made since 2014. "*Government policy had already started to encourage electro-mobility in 2001, when a key special project for EVs became part of the national high-tech R&D program. In 2009, the government identified EV production as one of several strategic emerging industries, creating a range of very attractive incentives and setting the ambitious targets of producing two million BEV/PHEV by 2020*" (Altenburg et al. 2015).

In July 2014, the Chinese government cut off a 10 % tax for drivers buying an electric or a hybrid car associated with a RMB 6000 allowance (about $1000). These incentives are also valuable for foreign cars. US$ 16 billion will be spent to build a charging station network for electric cars. This infrastructure development is also supported by foreign car makers such as Tesla, the American electric car maker, seeking for new outlets for its production. In 2014 T possesses 200 electric stations and agreed with China Unicom, a Chinese telecom company, to build 400 more stations.

In France, the government confirmed his willing to increase the number of electric cars on the roads mainly in urban areas and introduced a super allowance (€10.000) for the replacement of a car older than 13 years by an electric car. This allowance may imply more than 1.5 million cars. The French government aims at supporting electric cars' growth by strengthening restrictions and taxes on polluting cars.

The development of electric vehicles in France also leans on the modernization of the automotive sector, car makers and OEMs: the Fund for OEMs modernization (FMEA Fund) has been transformed into "*Fund Future car*", with €270 million to promote the emergence of new champions in the sector.

Similarly, the U.S. Government launched in 2009 a plan of $14 billion for the development of high technology cars including $8 billion for three car makers willing to produce electric vehicles and adapt their plants to this production (Ford, Nissan and Tesla Motors). Finally, the Recovery Act allowed unlocking an envelope of $2.4 billion for 48 R&D projects including a good part for OEMs.

In this context, the electric vehicle manufacturer Tesla has benefited from a large part since it has obtained more than $2.4 billion in grants since 2009, with tax credits for its batteries factory in Nevada ($1.3 billion), various assistance programs in California (storage of energy…), and also benefiting from tax incentives relating to clean vehicles, 7500–11,500 dollars according to the States, which represented an indirect subsidy of 284 million for the constructor on 2014.

Worldwide public incentives for the purchase of electric vehicle are numerous and are beginning to bear fruit with the beginning of real sales growth, although the market is still small with less than 500,000 registrations in the world in 2014. The interest in the electric vehicle should however grow progressively as drivers can try them in actual conditions on roads thanks to car sharing organization implemented by either public or private operators such as in France: Autolib car-sharing of electric cars in city services' (Paris), Bluely (Lyon), Bluecub (Bordeaux) or Autobleue (Nice) that promote mobility and inter-modality.

Leurent and Windisch (2011) analysis of public policies and projects in favor of electric mobility within an international review of national policies and regional projects showed that "*most countries consider explicitly a development path made up of three steps: from Initialization by pilot projects and procurement initiatives, to a Long Run step where EVs hold a large share of the car stock, passing by a Medium Run step which involves taxi and shared-car fleets*".

Leurent and Windisch (2013) detailed public policies implemented in 6 European countries (Germany, United Kingdom, Spain, Denmark, France, Norway), in the USA, Japan, China and India and the European Union as a whole. Their analysis underlines the fact that smart-mobility is fully embedded in larger "*derived challenges mainly: (a) the organization of mobility together with the uptake of recharge infrastructure, (b) the interaction of the electric mobility system with the electricity net, hereby considering the energy storage function of batteries,*

(c) the technological progress and the industrial production of vehicles and various components. Overall, these stakes involve the community more than the individual vehicle user".

The following second part of the chapter describes and analyses the specific case of public policies implemented in France, exploring how these challenges are being tackled by the French national and local authorities.

4.3 Public Policies in France: Toward a Sustainable Smart Mobility

Mobility in France is also experiencing a formidable transformation and governments, national or local, are major stakeholders in this transition, introducing a global approach of urban mobility.

In fact, «*a global approach allows the integration of technological innovations within the social context they interact with (…) and the conciliation of collective and individual choices in order to open the door to a sustainable and peaceful mobility*» (OPECST 2014). This will be possible thanks to the reinforcement of public policies in order to adapt collective choices to available resources and align economic, social and ecologic stakes. In the same time, governments must think how to associate citizens in the decision process and to propose new and more imaginative incentives (OPECST 2014).

The "*Automobile Pact*" released in February 2009, foresees a EUR 250 million loan for the industrialization of decarbonized vehicles. The "*Grand Emprunt*" (announced in December 2009) foresees EUR 750 million for the development of decarbonized vehicles. This funding will be invested in several research and deployment projects under the patronage of the French Environment and Energy Agency (ADEME). Specific funding has also been made available for the construction and development of a battery production factory with a capacity of up to 350,000 batteries. The eco-conception of batteries and their recycling are research priorities.

In order to assure the supply of appropriate recharging infrastructure, laws have been released that oblige every new building equipped with parking units to connect these to electricity supply by 2012. Car parks at work places have to be equipped with electricity connections by 2015. Further, EUR 60 million has been made available for the installation of 1250 public recharging points around 20 agglomerations till 2012. By 2025 a recharging infrastructure of 9.9 million points shall have been established around France (thereof 9 million private points, 750,000 public normal charging and 150,000 public rapid charging points) (Leurent and Windisch 2011).

But EV will not be the only vector toward a new mobility; a new organization of vehicles flows is being implemented, leveraged by digital technologies and a new way of thinking transportation. For the forthcoming decade at least, French policies will have to organize urban mobility combining four major dimensions:

- Traditional modes: private cars, two- and three-wheels vehicles, taxi service, public transit, walking
- On-demand mobility: car-sharing, car-pooling, bike-sharing, on-demand public transit
- Supporting infrastructures and services: smart parking, integrated traffic management, EVs charging systems, mapping service travelers information systems
- Technologies: data hub, connected vehicles, autonomous vehicles.

4.3.1 A Mobility Leveraged by Digital Technologies

The new approach of mobility is necessarily global, coupling offers and equipment from different actors (public and private) and relying on the tools and technologies of digital communication. Increasingly, the economic model of urban mobility is more and more based on information and telecommunication technologies supporting the multi modal framework.

This is the objective of *Optimod project*, developed by the City council of the city of Lyon in France: to implement a single browser, giving the user all mobility options based on multi-transport to go from point A to point B, on the basis of different criteria. It will integrate all functions associated with the GPS and will provide the user a real-time urban navigation tool. The collection of traffic data, in particular, gave rise to experimentation of innovative tools: wireless sensors that are gradually deployed in the city, tracer vehicles with anonymized positions allowing reconstituting all journey schedules. For this project, the Lyon council organized a partnership between private companies such as Orange (telecommunications), IBM (data mining) and Renault (vehicles) to create the browser.

As another example, the city of Saint Quentin-en-Yvelines in Paris region joined the French car maker Renault to launch a car sharing service for urban transfers working instantaneously with an app that localizes the closest vehicle to be rent.

This global approach to the urban mobility also becomes a reality within Saclay, south of Paris that regroups several clusters (Mov'éo, Systém@tic and Advancity) who implemented the SYSMO 2015 project aiming exploring new mobility's process. Carried by a pool of public and private companies (RATP, SNCF, Renault, Valeo, Continental), this project should build a union of all transportation options (public transportation, car sharing, carpooling, car or electric bike renting, all regrouped in a smartphone's app).

4.3.2 The Car in the Next Future: From Possession to Usage

As mentioned in the previous examples, the economic model of the multimodal dynamics more and more includes the existence of car sharing or carpooling that encourages a more sustainable and collaborative mobility.

On the French market, the number of organizations proposing to bring together cars' owners with non-regular drivers is growing (there are for example *Buzzcar*, *Greenie01*, *Deways*, *Koolicar*, *Livop*, *Mavoiturealouer.com*, *Voiturelib*). Different studies on car-sharing seem to converge to the following conclusion: the number of users will increase significantly over the next years (Ballet and Clavel 2007; Clavel et al. 2008).

The consulting company Navigant Research (2013) plans a phenomenal increase of the world market of the ride sharing, passing from 1 billion dollars in 2013 to more than 6 billion in 2020. The analyses of Ranke (2013) shows a similar progression, claiming that the number of users at the European level should indeed increase from 700,000 in 2013 to more than 15 million in 2020, and the number of shared vehicles from 20,000 to 240,000 over the same period.

The major French car manufacturers also entered the area, like the Citroën Company whose website comes to enrich new offers car-sharing oriented. These new services are car-sharing and carpooling (with covoiturage.fr). Starting from the premise that cars remain on average immobilized 90 % of the time, Citroën offers its customers to lend their vehicle when they do not use it. It is quite symbolic to see that an automaker, whose business model relies on the sale of cars, promote car sharing by definition opposed to buying car by individuals.

Another form of car-sharing, i.e. carpool-pooling, continues to make progress thanks to the rise of social networks and smartphones' applications. Indeed, thanks to the emergence of "*dynamic car-sharing*" (also called "*real-time car-sharing*"), greater flexibility and greater speed of communication allow to offer instant car-sharing. 10 % of U.S. citizens use carpool, a figure that is expected to triple by 2020. The French company *Blablacar*, created in 2006, recorded 10 million users in 2014 and is expanding worldwide.

Indeed, the emergence of the "*dynamic ride sharing*" allows both a better flexibility and a higher speed of mobility. This dynamic ride sharing is also boosted by the democratization of smartphones and their applications offering geo-localization for a quick getting in touch. Chassignet (2014) shows that there are in reality several car sharing's types, among which the "*direct track*" allows the customer to let the car in any station or even on the public road (for example, *Autolib'* in Paris) or the "*user-to-user*" sharing organized on a web community platform managing direct links between car owners and car users (for example: *Buzzcar*, *Drivy*, *Ouicar*).

For sure, all these offers wouldn't have encountered a so quick success if not supported by public policy, not only with the adaptation of regulations but also, as we presented it on previous examples, with large implication of local governments in the evolution of urban mobility.

4.3.3 The Role of Public Policies in the Emergence of the Autonomous Car

In France, economics and politics are evolving within a centralized state; this centralism should logically lead to a massive and active intervention of the national and local governments in order to meet the requirements of smart mobility. But in practice, decision process and actions need quite a long time to be taken and implemented. But as far as now, no regulation has been imposed. The national government has largely transferred the investment in infrastructures to local authorities. Consequently, many local initiatives have been emerging in place of national regulation.

We can observe a paradox within the two French car makers, Renault and PSA Group (Peugeot, Citroen). The State is a shareholder of the two companies but it apparently did not influence the design and production of an electric car in France nor within the two companies alone or in partnership. This situation underlines the permanent ambiguity between state directives on modern mobility and private strategies that should *walk the talk.* This distortion has slowed down the development of a sustainable modern urban mobility.

However, the new ecosystem of modern mobility leans on an innovation whose technologies are establishing a significant disruption: the autonomous vehicle. This automobile revolution forces the government to act differently.

A study of IHS Automotive (2014) considers that the fully self-driving car is not waited before 2025 but vehicles with automated-driving will represent about 12 % of the world sales in 2035 underlining besides that "*the number of accidents will be close to zero for these autonomous cars, the road traffic could be then regulated and the air pollution mastered thanks to the development of programs optimizing the energy consumption*".

French authorities do not doubt that the autonomous car will transform the car industry. The CEO of the French car manufacturer Renault claims that two thirds of Renault's sales amount planned for 2025 will indeed be related to software, on-line data processing and security systems. The car itself, produced by the manufacturers and OEM, will only represent one third of value. Therefor car manufacturers must worry with the arrival on the market of new comers like Google, Tesla or Apple that may be already very advanced in their programs.

Car manufacturers must react quickly. So did Nissan (Group Renault) that managed to have recorded in Japan its Leaf, a car equipped with a large automatic driver's assistance.

Donada and Fournier (2014) analyzed the arrival of these big companies whose board have considered the new mobility as a source of strategic diversification (Samsung, BYD, Bolloré). With the same vision new comers are appearing today like giants of different markets such as IT sector (IBM, Google, Apple), energy or infrastructures (EDF, Air Liquid, Vinci, Veolia), networks open-source (Wikispeed), mixed structures of laboratories and companies (InEco project between researchers of Dresden and ThyssenKrupp) and public services (La Poste, ERDF).

A reorganization of activities is already planned for French car makers that have to think about cooperation and partnership with new comers. But cooperation is complex and the question to know if users will quickly accept this disruptive innovation for their mobility is still asked.

This rapid arrival of new comers has induced not only a big change in the services' offer to urban mobility but also important modifications in users'behaviors and manufacturers' strategies. Thereafter, the ecosystem of mobility is being reshaped; roles and the powers of the various stakeholders are evolving, giving birth to a new industry, the mobility 2.0.

French governments are rather positive, believing that the deployment of autonomous cars and all new forms of mobility, will help to smooth and safe urban traffic and reduce pollution. Nonetheless, this trend raises new issues, mainly the need to rethink the legal framework. Indeed the Vienna Convention, signed in 1968 by most European countries, specifies that the driver is solely responsible at any time for the behavior of his vehicle. But in an accident with a self-driving car, the question of responsibility isn't so clear. What about the responsibility of public authorities that organized the transport policy and the one of the supplier that provided embedded technology or the manufacturer that integrates it to the car?

Nevertheless the French government is pushing French car manufacturers not to fall behind and to tackle the point to remain in the race—one must not forget that the French government is a shareholder of the two French car manufacturers Renault and Peugeot.

Strong partnership strategy has been deployed by the two companies. As an example, the Science Technologies Exploratory Lean Laboratory (StelLab) of PSA Peugeot Citroën was founded in 2010 to oversee scientific partnerships and coordinate the group's research OpenLabs. StelLab is responsible for leading an interdisciplinary network that fosters discussion and dialogue among scientists and experts from the mobility industry. Its task is to identify and develop the new technologies and innovative business models of the vehicle of the future. StelLab is a very good example of public and private partnership that may be developed in France to tackle complex issues combining as well technological challenges as sociological trends and economical stakes.

Yet, fleets of autonomous car are not to be seen right away on the roads; prospective analysis forecasts progressive steps from traditional car to autonomous cars distinguishing 5 levels from full driver responsibility to full vehicle responsibility:

- Level 1: vehicle provides drivers information warnings
- Level 2: vehicle integrates detection and response
- Level 3: vehicle is autonomous, driver takes control in emergency
- Level 4: vehicle fully autonomous, occupant do not need ability to drive
- Level 5: vehicle fully autonomous, connected and cooperating, optimized system operations and passive occupant experience.

Levels 4 and 5 are considered by French authorities as really disruptive and in the next future, French urban spaces will quite surely implement these last two levels on dedicated areas or lanes.

4.4 Conclusion

The development of a new mobility will not only cause a complete disruption of urban mobility services for users, but it will turn the table for all stakeholders evolving in the car industry (manufacturers, energy companies, computer scientists and public authorities).

Among these stakeholders, the public actor (i.e. local and national governments) plays a leading role in the world in transforming the automotive industry into the ecosystem of smart-mobility 2.0. The States, the European community as well as local authorities are implementing urban mobility policies imposing global guidelines for a sustainable responsible mobility framework. In fact, the real point is that "*the new mobility is questioning politicians*" (Donada 2014), and governments have to remain watchful and active in order to face the complexity of this new ecosystem of mobility, an open and multidimensional ecosystem whose borders exceed widely those of the car industry.

In the large and medium-sized cities of the world, various innovative actions at national or local level are carried out by the governments, in order to lead the evolution towards the new mobility. These actions aim at significantly improving quality of living for urban inhabitants.

But governments have to define the outlines of a sustainable and responsible mobility to use effectively the levers of the economic development for large cities that are still enlarging and growing. Politics have to reshape urban landscapes to encourage a new multimodal, connected and autonomous mobility. This will be the public challenge for the very next future. In that respect, governments are playing the game as well with incentives to boost as with constraints to guide towards the sustainable development for a new urban mobility.

Acknowledgments This research benefits from the support of the Chair "PSA Peugeot Citroen Automobile: Hybrid technologies and Economy of Electromobility", so-called Armand Peugeot Chair led by CentraleSupélec, ESSEC and sponsored by PEUGEOT CITROEN Automobile.

Appendix

See Tables 4.1 and 4.2.

Table 4.1 Various measures implemented to resize transport in urban areas

Examples of implemented regulations		
Standards of CO_2 emissions	Only vehicles proving a certain level of CO_2 emissions can be sold	EU: CAFE regulation (clean air for Europe)
Fuel standards	Only fuels with low CO_2 emissions can be sold	USA: goals on biofuels (15 % in 2015)
Speed limits	Establish the most appropriate speed limits	Spain: 30 km/h in urban areas
Low emissions zones (LEZ)	Limited access in some areas to only low-emission vehicles	Tokyo: LEZ 2003 Berlin: LEZ 2008
Circulation with alternating traffic	Limited access to certain vehicles according to their car plate number	Beijing: during the 2008 Olympic Games
Restriction on the possession of the vehicle	Restrictions on ownership of the vehicle through a quota system	Shanghai: clean vehicles entitled to free registration
Parking restrictions	Limit the duration of parking options	London: parking allowed between 8:30 and 6:30 p.m.
Infrastructure		
Reserved lanes	Creating dedicated lanes for vehicles with high occupancy rates, permanently or not	Beijing: dedicated lanes for buses during peak hours
Soft modes of transportation	Creating dedicated lanes for pedestrians, bicycle or clean moto cycles	Australia: promoting cycling
Price control		
Review of charges and subsidies for car purchase	Review of taxes or subsidies on new vehicles, low taxation, purchase subsidies	China: 10 % increase in the car price except for clean vehicles
Taxes, subsidies and grants on use	Emission taxes, fuel taxes, tax on kilometers traveled, tolls, parking fees, at kilometer Insurance	London: downtown toll
Pricing public transport	Free Pricing, zonal pricing, progressive rating	Shanghai progressive pricing

(continued)

Table 4.1 (continued)

Volume control		
Quotas market	Limit the amount of CO_2 emitted per year and afford to buy or sell CO_2 allowances or quotas	EU ETS (Emissions Trading System) trading system
Equipment and technologies		
Optimization of vehicles	Optimized vehicles (hybrid, electric)	China: promoting vehicle electrical
Intelligent Traffic System (ITS)	To monitor and regulate traffic and provide data that could generate new business	Amsterdam: ASC (Amsterdam Smart City)
Public transport development	BHSL (Bus with High Service Level), tram, subway, train, boat	Brazil Curitiba BRT (Bus Rapid Transit) or BRTHSL (Bus Rapid High Service Level)

Source Michelin (2014)

Table 4.2 Examples of politics implemented in cities to reduce traffic's nuisances

London	Beijing	Los Angeles
Key traffic measures: **Congestion Charge Zone**: Fee charged on most motor vehicles operating in Central London (exemption is linked to CO_2 emissions for cars that emit >100 g/km) **Low-Emission Zone**: The zone covers most of Greater London. Truck, bus, coach or other specialist heavy diesel vehicles in the Low-Emission Zone (LEZ) need to meet certain emissions standards (cars and motorcycles are not affected)	*Key traffic measures*: **Odd-even license plate system**: Allows cars to drive on alternate days, based on the license plate number (fine 100 yuan) **Traffic restriction for non-residents of Beijing**. **Increasing parking fees**. **Control of license plates** (monthly plate lottery).In 2011 announced the introduction of Congestion Charge—not implemented yet	*Key traffic measures*: **High Occupancy Vehicle lanes (HOV)**: Restricted traffic lanes for exclusive use of vehicles with a driver and one or more passengers **High Occupancy Toll (HOT)**: Congestion pricing that gives motorists in single-occupant vehicles access to HOV **ATSAC**: (Automated Traffic Surveillance and Control) **Plans to improve public transport**

Source Little (2014)

References

ADEME, 2014. Les zones à faibles émissions à travers l'Europe : Déploiement, retours d'expériences, évaluation d'impacts et efficacité du système. Agence de l'Environnement et de la Maîtrise de l'Energie.

Altenburg, T., Schamp, E.W., Chaudhary, A., 2015. The emergence of electromobility: Comparing technological pathways in France, Germany, China and India. Science and Public Policy scv054. doi:10.1093/scipol/scv054

Arthur D Little, 2014. The Automotive CO2 Emissions Challenge.

Ballet, J.-C., Clavel, R., 2007. Le covoiturage en France – État des lieux et perspectives, Certu.

Buchan, D., 2012. he Energiewende – Germany's gamble. Oxford Institute for Energy Studies, Oxford, UK.
Chassignet, M., 2014. Étude sur les pratiques d'autopartage et de covoiturage. Les Transports du futur.
Chriqui, V., 2010. Les nouvelles mobilités, Rapport du Centre D'Analyse Stratégique, La Documentation française.
Clavel, R., Mariotto, M., Arsac, B., 2008. «L'autopartage en France et en Europe, Certu.
Delft, C., 2011. Impacts of Electric Vehicles. April, online accessible via www. cedelft. eu (20-09-2011).
Donada, C., 2014. Les sciences de gestion, boussole du politique. Étude sur l'émergence d'une nouvelle industrie de l'électromobilité. Revue française de gestion 40, 35–54.
Donada, C., 2013. Electric mobility calls for new strategic tools and paradigm for automakers. International Journal of Automotive Technology and Management 13, 167–182.
Donada, C., Fournier, G., 2014. Stratégie industrielle pour un écosystème en émergence: le cas de la mobilité 2.0, décarbonée, intermodale et collaborative. Revue d'économie industrielle 317–348.
Fréry, F., 2000. Un cas d'amnésie stratégique: l'éternelle émergence de la voiture électrique. Presented at the IXème Conférence Internationale de Management Stratégique. Montpellier.
Freyssenet, M., 2015. La production mondiale des 50 premiers constructeurs automobiles existants, 1945-2014. Document d'enquête: quatre tableaux et six graphiques commentés. Édition numérique, freyssenet. com, 2007 et mises à jour.
IHS Automotive, 2014. Emerging Technologies: Autonomous Cars-Not If, But When.
Jullien, B., Pardi, T., 2013. Structuring new automotive industries, restructuring old automotive industries and the new geopolitics of the global automotive sector. International Journal of Automotive Technology and Management 20 13, 96–113.
Leurent, F., Windisch, E., 2013. Politiques publiques et déploiement des véhicules électriques. Tec 220, pp-34.
Leurent, F., Windisch, E., 2011. Triggering the development of electric mobility: a review of public policies. Eur. Transp. Res. Rev. 3, 221–235.
Michelin, 2014. Michelin Green Paper, Dare innovative mobility for urban growth and welfare.
Navigant Research, 2016a. Changing Models for Urban Mobility.
Navigant Research, 2016b. Carsharing Programs.
Navigant Research, 2013. Carsharing Programs.
OPECST, 2014. Les nouvelles mobilités sereines et durables : concevoir et utiliser des véhicules écologiques, Publication Assemblée Nationale, N°1713 et Sénat, N°293 sous la Direction de D. Baupin, Député et F. Keller, Sénatrice.
Ranke, M.L., 2013. Car Sharing End-user Analysis in Selected European Cities. Frost & Sullivan.

Part II
A Mobility Revolution at the Dawn of a Societal Revolution

Chapter 5
Extending the Scope of Partnerships in the Automotive Industry Between Competition and Cooperation

Danielle Attias and Sylvie Mira-Bonnardel

Abstract Faced with a drastically modified social and economic context and a highly competitive global market, the automotive industry must redefine its strategy but is struggling to find the *right* positioning. While cooperation with traditional players is necessary, carmakers find themselves obliged to form alliances with new entrants, often far removed from their core business. The automotive sector currently faces numerous challenges, i.e. master and develop new technologies to respond to new usages, and ensure steady production volumes to make their activity long-lasting, while adapting to an uncertain, globalized economic world. In this situation, an analysis of carmaker strategies should consider the mobility ecosystem, which results in new partnership approaches. The scope of partnerships is widening, the automotive industry's *former* production models are no longer suitable and new economic models are emerging. Car manufacturers' traditional strategic choices, fluctuating between competition and cooperation, are already outmoded because the automotive sector needs to anticipate a future ecosystem whose epicenter will be the *intelligent* and autonomous car. In this respect, the challenge of capturing value remains crucial and worrisome, in particular when most profits are made from mobility services. A reorganization of all players in the industry is currently taking place, and one of the features of this transformation is the emergence of *coopetition* strategies, presented in this chapter. This is a new era in which value creation fits not just into the relationship dynamics between partners in innovation ecosystems, but also into a system jointly built with other allies, from outside the automotive industry. Will this interweaving of companies be profitable for all in the long run?

D. Attias (✉)
Industrial Engineering Research Department/Laboratoire Génie Industriel, CentraleSupélec, Université Paris-Saclay, Grande Voie des Vignes, 92290 Chatenay-Malabry, France
e-mail: danielle.attias@ecp.fr

S. Mira-Bonnardel
Ecole Centrale Lyon, 36 Avenue Guy de Collongue, Ecully 69134, France
e-mail: sylvie.mira-bonnardel@ec-lyon.fr

D. Attias (ed.), *The Automobile Revolution*, DOI 10.1007/978-3-319-45838-0_5

Keywords New business models of automotive industry · Coopetition between partners · Sustainability ecosystem of mobility · Strategy of value creation

5.1 Introduction

According to a report by KPMG[1] (2010), 68 % of automotive industry majors favor developing innovation within strategic alliances or through creating joint ventures between automobile manufacturers and parts manufacturers. The transformations taking place in the industry have seen partnerships develop and diversify, in a constantly moving ecosystem involving multiple stakeholders, such as suppliers, distributors, telecommunications operators, digital firms, local authorities, regulatory authorities and competitors. Increasingly, automobile manufacturers are taking a competition-cooperation approach, coined as "coopetition" by Brandenburger and Nalebuff (1996).

Coopetition brings manufacturers both the benefits of competition, such as stimulating and mobilizing teams, and the advantages of cooperation, such as access to resources, skills, and broader markets (Bengtsson and Kock 2000; Brandenburger and Nalebuff 1996).

The coopetition strategies that emerged in high-tech industries (Apple and Samsung, Sony and Ericson) have now extended to the automotive industry (Le Roy et al. 2010). One example is the *Global Hybrid Cooperation*, a coopetition agreement between General Motors, Daimler/Chrysler and BMW, to develop a common hybrid system, or the *Hybrid Synergy Drive*, a coopetition agreement signed by Toyota and the Renault/Nissan group.

Today, successful coopetitions are based on strategic communities commonly known as "business ecosystems" (Moore 1993) or "innovation ecosystems" (Iyer and Davenport 2008). Although analyses concur that the benefits of cooperation within strategic communities are driving factors in product/service innovations through cross-fertilization (Jullien 2000), their implementation is nevertheless still restricted by both the politico-economic context and the strategic objectives of each stakeholder company in terms of capturing the value created.

Thus, the "*global strategic alliance*" announced between the groups PSA Peugeot Citroën and General Motors came up against a set of constraints that prevented the cooperation from lasting. Numerous other coopetition experiences show that successfully implementing common synergies remains problematic and is strongly influenced by the contexts in which companies operate.

In fact, confronted with an environment going through significant technological and societal change, the leaders of major industrial companies in the automotive industry are obliged to make totally different strategic choices in order to remain

[1]KPMG study (2010) of the 200 leading automobile manufacturers, parts manufacturers and distributors in the world.

prime players in their innovation ecosystems, and thus capture a significant share of the value. These choices lead them to extend the scope of their partnership strategy, as we demonstrate in this chapter.

This chapter is organized into two parts. In the first part, we present the different theoretical studies that have led to a better understanding of how the strategic contexts have changed with the growth of organizational interaction. The second part describes alliance configurations and how they are evolving in the new strategic contexts of the automotive industry. A study of the automobile group PSA Peugeot-Citroën serves to illustrate this issue.

5.2 The Automotive Industry: A Durable Innovation Ecosystem

5.2.1 From Automotive Industry to Mobility Ecosystem

As early as 1986, Thorelli, in a seminal article entitled, "*Networks: between Markets and Hierarchies*", pointed out the benefits of networked structures, positioning them as forms of governance halfway between market and hierarchy (Thorelli 1986). In comparison to standard set-ups, networked organizations have less formal, less specialized structures and vaguer borders, making them better suited to modern competitiveness conditions. Their flexibility and reactivity led Livian (1998) to remark that a networked organization is "*a new ideal-type*", echoing Fréry (1997), who esteemed that, "*The major integrated capitalist company will have simply been an episode in history (…) an episode that seems to be over.*"

The supremacy of networks as modes of governance became even stronger with the reorganization of value chains and the externalization of activities to boost companies' flexibility and agility (Rugman and D'Cruz 2000). Similarly, networks quickly came to be seen as the best organizations to create the conditions for the emergence of a collective intelligence, to form learning hubs and to encourage innovation (Julien et al. 1996; Rogers 2003).

At the same time, Moore (1993) took another theoretical approach to inter-organizational relations when he introduced the concept of a business ecosystem based on a biological metaphor in two significant publications: "*Predators and Preys: a new Ecology of Competition*" in 1993, for which he won the *McKinsey Award* for the best article, followed by the book, "*The Death of Competition*" in 1996.

Most publications on business ecosystems give more or less the same definition, i.e. "*An economic community supported by a foundation of interacting organizations and individuals—the organisms of the business world. The economic community produces goods and services of value to customers, who are themselves members of the ecosystem. The member organisms also include suppliers, lead producers, competitors, and other stakeholders. Over time, they coevolve their capabilities and roles, and tend to align themselves with the directions set by one or more central companies. Those companies holding leadership roles may change over time, but the function of ecosystem leader is valued by the community because it enables members to move toward shared visions to align their investments, and to find mutually supportive roles*" (Moore 1996, p. 26).

The concept of an ecosystem is therefore a reference to an extended environment of coopetitive partnerships in which heterogeneous stakeholders with specific skills are likely to intervene to various degrees in a process of collective value creation piloted by only one of them.

Consequently, the strategies employed within business ecosystems differ depending on the position of their stakeholders. Iansiti and Levien (2004) identify four stakeholder categories, i.e. "*niche players*" that specialize in specific expertise; "*dominators*", whose main goal is to control the network; "*hub landlords*", go-betweens who encourage expansion and manage communication nodes; and "*keystones*", the leader companies that orchestrate cooperations and "*have to constantly manage the tension between creating value and capturing the value created*" (Isckia 2010).

Numerous analyses recognize the benefits of business ecosystems in terms of capital gain and innovation capacity for companies (Iansiti and Levien 2004). In their work on electromobility, Donada and Attais (2013, 2015), then Donada and Fournier (2014), showed that the automotive industry has entered a new era in which value creation fully depends on the dynamics of relationships in an open innovation ecosystem with mobile borders on a long-term grid. These authors clearly set out the complexity of organizing and managing innovation processes in such a context and the need to therefore conceive new, more suitable, innovative modes of governance.

The automotive industry is perfectly characteristic of these ecosystemic dynamics, the perimeter of which has extended considerably over the last decade. Rishi et al. (2008) include in the automobile ecosystem sectors as varied as, "*energy, consumer electronics, communities, geographies, social networks, other industries (software, telecommunications, financial services), and government*".

Dammenhain and Ulmer (2012) show that the automotive industry's transition towards an electromobility ecosystem has taken partnership strategies way beyond the automotive industry, "*The e-mobility ecosystem comprises vehicle manufacturers, suppliers to them, the electric vehicles, the IT provider, the e-mobility technology supplier, the e-mobility provider, the public sector, the utility, the distributor, and the charging/changing operator.*"

To take part in the electromobility era, the automotive industry is obliged to redefine its product range to offer a combined *product-service system* (PSS) and even a *modular product-service system* (MPSS). As a result, traditional players in the industry find themselves obliged to form new alliances with companies in emerging sectors (e.g. performance economy, circular economy, digital economy, etc.): "*There is a dichotomy between on the one side companies designing and manufacturing cars, and on the other side, companies operating result-oriented services. (…) Yet, an important part of the opportunities offered by PSS lies on the correlation between product and service activities, which is not totally achieved with the dichotomy mentioned*" (Mahut et al. 2015).

When innovation development fits into an open innovation rationale (Chesbrough 2006), new strategies emerge, resulting from a combination of industrial property, value creation and market differentiation strategies, and *in fine*, strategies for capturing the created value. In fact, although coopetition brings the numerous advantages of cooperation, it also involves multiple risks (Park and Russo 1996), including an imbalanced capturing of created value.

5.2.2 The Business Ecosystem, a Hotbed for New Mobility Innovation

The automotive industry is gradually transforming into a new ecosystem, i.e. the ecosystem of intelligent mobility, and this mutation is destabilizing secular collaborations between powerful parts manufacturers and automobile manufacturers. The growth in electric cars sales represents a significant drop in value for OEMs (no more thermal combustion engines, simplified drivetrains, etc.), which have no strategic advantage in developing electric vehicles and need to establish a new strategic positioning in the face of manufacturers' more wide-ranging partnership strategies: from university research centers (Barnes 2002) to local authorities, and industrials from other sectors such as telecommunications and IT.

Sarasini (2014) undertook extensive bibliometric research on patents filed in new electric vehicle technologies, highlighting the importance of collaborations between automobile manufacturers and academic research institutions: "*Academic institutions are key partners for automakers. It may be the case that academia acts as a supplier of skilled labor to the automotive industry, as a source of ideas and/or as a site for experimentation. Academia is perhaps even an indirect source of innovation for automakers even though this is difficult to evaluate using patent records*".

The mobility ecosystem constitutes a hotbed to the extent that it could easily be qualified as an innovation ecosystem (Iyer and Davenport 2008; Miller et al. 2008) inspired by Google's innovation ecosystem model. In this system, Google plays the role of "keystone firm" (Iansiti and Levien 2004) for an innovation capacity multiplied by the n stakeholders in the ecosystem, each of which generates an auto-captured value that is as beneficial to the ecosystem as it is to its leader.

An understanding of Google's ecosystemic strategy is crucial for automakers. Indeed, the question of the autonomous connected car and the new but primary role that Google now plays in the car industry constitute the keystone of new strategies. Clearly, Google has set out to revolutionize the mobility industry. On the one side, the company masters 3D software and mapping, in which it has made huge investments with the deployment of Google Maps and its recent buy-out of Skybox Imaging, the only company in the world that possesses several 3A imagery satellites. On the other side, the American internet giant is seeking to revolutionize mobility with its ambitious project of an economic model offering freely available autonomous cars in a car pool.

If this strategy is implemented in the coming decades, it will be destructive for the automobile sector because the American giant easily possesses sufficient funds to invest in industrializing its project on a large scale. Google's announcement that its 25 Google Cars have already covered one million kilometers proved its competitive advantage (Cazenave 2014).

Google's strategy, although embedded in a broad network of partners, raises questions and tensions: "*Ecosystem-oriented innovators strive to avoid the appearance of competition by claiming to help everyone. For example, Google executives seldom miss an opportunity to remind the world that they don't compete with media and content companies. Instead, they characterize media companies as their partners. Not everyone is so sure*" (Iyer and Davenport 2008).

A study by Ramirez-Portilla et al. (2014) of the ten biggest automobile manufacturers in the world[2] shows that these companies have all considerably boosted their open innovation practices since 2005, developing multiple partnerships to innovate. Firstly, by forging vertical connections between customers and suppliers that are particularly prolific innovators in the car sector, where parts manufacturers have for a long time been partners with high innovation capacity; then bolstering these alliances with horizontal connections with competitors (GM and PSA alliance, Renault and Fiat alliance, PSA and BMW alliance); lastly, by extending the partnership scope to take in complementary companies (like Tesla or Google, and Spotify, which has partnered with Ford).

Studies of Italian automobile manufacturers (Di Minin et al. 2010) and German carmakers (Ili et al. 2010) point to the significant and positive impact of this practice on the economic performance of the companies analyzed.

The automobile ecosystem is therefore organized around different stakeholders that interact on a partial convergence of interests and objectives according to their own specific strategy. These rapprochements are not devoid of friction. Although the cooperation is guided by the need to work together for much greater value creation, the capturing of the value created leads to significant tensions that can jeopardize the pursuit of the cooperation (Bengtsson and Kock 2000).

In fact, working in a partnership within the ecosystem does not boil down to naïve optimism because competing companies are most often seeking to take control of the system, thus generating significant tensions—for example between Google and Microsoft, Apple and Samsung, and Microsoft and Linux.

Companies have entered into coopetition areas (Brandenburger and Nalebuff 1996) combining both cooperation and competition in a configuration that goes from simple or complex dyadic coopetition to coopetition in a complex network (Le Roy et al. 2010). This new partnership deal requires rethinking set-ups, organizations and economic models. However, the partnership relationship between competitors always raises the same question of control and leadership within the

[2]The carmakers studied were Ford, Toyota, Renault, Scandia, Daimler, Hyundai, Dong Feng Motor, Tata, BMW and Fiat.

cooperation, against a background of regulating the balance between competition and cooperation (Word 2009).

5.2.3 Reworking the Economic Models of the Automotive Industry for the Responsible Development of the Entire Ecosystem

The need to review the automotive industry's economic models in line with responsible development does not date from the 21st century. Back in 1990, Nicolas Hayek declared his intention to create the Swatchmobile based on interdependent trends that he thought indicated a change in the car industry: integrate an ecological dimension, resolve urban journey problems, move from "ownership to hire" and create added value on services (Métais and Pin 2002). "*The idea behind the Swatchmobile project is based on a simple observation: in a built-up area, a car travels an average of 30 km per day, carries 1.2 passengers and remains stationary for over 90 % of the day. During the remaining 10 %, the backseats are usually taken up by the newspaper. Generously, we have built a car for two*" (Métais and Pin 2002).

Ahead of its time, the Swatchmobile did not manage to replace the old model with a new conception of urban mobility. The need to strike an alliance with a carmaker led Mercedes to take over the concept and transform Hayek's vision into a "car", despite the fact that his radical aim was to make the Smart anything but a car.

Could this new concept of mobility and the role of the car become the rule in the very near future, while carmakers remain caught up in an inflexible production system?

Two factors argue in favor of this transformation, i.e. the massive arrival of new entrants free from technological and industrial constraints, and societal changes associated with new public policies.

The first driver of change—the arrival of new entrants—obliges carmakers to profoundly reconsider their strategy for the coming decade. These new entrants are of course Tesla and Google, as well as less well-known Chinese industrials like BYD, whose European director confirmed his intention to open at least one production plant on the continent at the 2015 Customer Experience Summit.

BYD specializes in batteries for cell phones and is the world leader in Lithium-Ion accumulators, with 30 % of the market. Established in 1995, the industrial started diversifying in 2003 when it invested in the car sector. This daring move, combined with its battery expertise, made the young automobile manufacturer one of the leading players in electric mobility in China. Its catalogue includes electric buses and individual electric and hybrid rechargeable cars. Among the latter, the Qin sedan and the Tang Plug-in Hybrid SUV are due to arrive on the European market in 2017. In partnership with Daimler, BYD now offers an electric

Denza sedan with 300 km of autonomy using a battery with a capacity of 47.5 kWh.

The second driver of change results from societal changes and the many public policies championing a new mobility, as discussed in Chap. 3. These currently constitute significant groundswells that carmakers cannot ride today as they did at the end of the 20th century.

Without doubt, automobile manufacturers are at the crossroads of two major changes: the transition towards the ecosystem of intelligent mobility, and the performance economy. These two major trends aim at the implementation of sustainable, responsible economic models. Since sustainable development questions are particularly aimed at companies in the automotive industry, carmakers have understood the need to organize a global response involving multiple collaborations within the entire mobility ecosystem, even if that means offering products that compete head on with their car sales, such as Citroën, which sells, hires and shares its vehicles.

The development of collaborations within an extended perimeter also leads carmakers to cooperate with their competitors, establishing the coopetition situation mentioned above, which is based on a technico-economic rationale of pooling resources, considering that, "The *core problem for the automotive industry is that it is insufficiently profitable, particularly given the capital intensity of the industry*" (Nieuwenhuis and Wells 2003). Developing an extensive partnership strategy thus becomes one of the foundations of the economic model. However, implementing a responsible, sustainable economic model requires going a lot further in terms of pooling and generalizing the developments of the industry.

Car industrials are obliged to develop a sustainable economic model, in other words: "*A sustainable automotive industry is one that creates life-enhancing employment for communities over a long period of time. It has zero net consumption of physical resources in production. It is consistently profitable while being able to withstand short-term fluctuations in economic circumstances. And it produces products that themselves do not pollute or otherwise degrade the environment, are fit for purpose, and are designed for longevity. All of these features suggest that, over time, manufacturing as such (of new, complete products) would become only a small part of the business model, that concepts such as product-service systems (PSS) are more appropriate. It is recognized that profitability is absolutely vital for sustainability. However, profitability is a necessary but insufficient condition for sustainability: the environment and social dimensions must also be included*" (Wells 2013).

This new economic model connects the fourfold product-service-structure-market to the new technologies of electro mobility and to the societal shift towards a sharing, circular, usage, etc. economy for which digital technologies permit the development and open the way to responsible, sustainable mobility.

The question of partnership strategy in an ecosystem governed by long-lasting, sustainable and responsible development objectives (Wells and Nieuwenhuis 2012) thus makes perfect sense because no industrials in the industry can drastically modify their economic model without impacting the industry as a whole.

Following a century of linear development, the automotive industry is entering a period of profound change and renewal of its economic models for sustainable, responsible insertion into an ecosystem in transition (Christensen et al. 2012). We give more specific details in the next part on the strategies of French carmakers.

5.3 French Car Manufacturers' Strategic Choices

5.3.1 Strategic Alliances for Innovation: Between Opportunities and Constraints

Among the successful automobile alliances, the Renault-Nissan partnership is often quoted as an efficient, long-lasting *coopetition* model. The alliance started in 1999 in the face of general skepticism, but has prospered and remains solid, while other cooperation set-ups failed during the same period. Examples are Daimler and Chrysler, Fiat and Chrysler, then Fiat and General-Motors, Volkswagen and Suzuki—the list of attempts of unions or partnerships that ended in failure is long.

When asked about the success of the Renault-Nissan cooperation, Carlos Ghosn, currently CEO of both manufacturers and the builder of this alliance, indicated several keys to success in constructing a long-lasting cooperation-competition arrangement.[3] For him, mutual trust between partners is crucial, and the merger-acquisition set-up that implies a strong partner preying over a weak one is an obstacle. The meeting of personalities with the desire and the will to successfully move closer together largely contributed to the coopetition's success.

One point should be underlined: the success of the alliance is mostly due to the respect for cultural differences and a consideration of the respective identities of a French carmaker and a Japanese one. In their book, *Citoyen du monde*, Ghosn and Riès (2005) defend this complementary cultural approach that made it possible to jointly build a common identity. The director of Renault also insists on a mutual respect for culture and *a fortiori* management modes that can be employed differently in each unit of the alliance. This multicultural management style is coupled with a project management style that is *innovative in its methods*. Starting from 2005, C. Ghosn encouraged commitment and consensus from all of those concerned in the company, whom he considered as essential components of success (Magee 2005). This commitment should be made concrete and communicated at both Nissan and Renault. The head of the company should be an example and show just as much commitment as the employees.[4]

In addition, due to precaution, intuition or pure strategic rationality, the decision was taken to implement the alliance stage by stage. Thus, for Renault-Nissan, the first phase of the strategy consisted in splitting global markets fairly

[3]Journal *l'Usine Nouvelle*, 28 February 2012.

[4]"Portrait d'un communicant", Journal *Stratégies*, 21 April 2005.

(Renault remained in Europe and Latin America, with Nissan taking Asia and the USA), then in developing strong synergies on common purchases, shared platforms, parts, and techniques, followed by an extension to logistics and marketing. Fifteen years later, it is indisputably the longest-lasting alliance in the automotive industry.

5.3.2 *The Case of the Automobile Manufacturer Peugeot-Citroën*

During the five years that we have been responsible for the Armand Peugeot Research Chair of the PSA Peugeot Citroën group, we have worked closely with the strategy of this major manufacturer. The Armand Peugeot Research Chair is part of PSA's StelLab. Created in 2010, the StelLab (Science Technologies Exploratory Lean Laboratory) is a structure that organizes scientific partnerships with the aim of coordinating research OpenLabs and creating an interdisciplinary network of exchange and dialogue between scientists and experts from PSA Peugeot Citroën. Its ambition: to identify and develop new technologies and innovative business models for the vehicle of the future.

This automobile group has an unusual history. The PSA group was involved in innovation early on Loubet (2004), and in the early 2010s started working on electric cars in partnership with Mitsubishi. It also produced two diesel/electric hybrids using *Hybrid4* technology that were presented at the Paris Motor Show. PSA's three-cylinder 1.2 PureTech engine won the 2015 International Engine of the Year award in its category and came second across all categories, beating dozens of competitors.

In June 2015, PSA Peugeot Citroën did not hesitate to change direction and announced a joint enterprise with the group Bolloré to create a carpool of electric or low-CO_2 emission vehicles, and cooperate in the electric vehicle domain.

This history of this carmaker features numerous cooperation's, some of them successful, some more problematic (Fernandez and Le Roy 2013). We worked with the group's scientific managers, who gave us details on the different alliances, such as the cooperation with General Motors and the alliance between PSA Peugeot Citroën and IBM.

5.3.2.1 Alliance Between General Motors and PSA Peugeot-Citroën, the Lessons of a Failed Partnership

The "*global strategic alliance*" established in 2012 between General Motors and PSA Peugeot-Citroën was announced as a Franco-American strategic cooperation on common programs to develop vehicles, and prefigured a reorganization of the global automotive industry. Clearly, an alliance between the American market leader and a French main player was likely to upset the European car market and, by extension, force global carmakers into a different positioning. The initial idea was very similar to that of the Renault-Nissan alliance: it involved pooling the crossed manufacture of vehicles and grouping purchases and logistics. A previous attempt at a major strategic alliance between PSA and the Japanese firm Mitsubishi had failed, and this new alliance was eagerly expected to create a new commercial synergy and financial gains for the French carmaker. One and half years later, the alliance was put to one side to make way for another rapprochement between PSA and the Chinese group Dong Feng.[5]

How should this alliance strategy be analyzed? Firstly, the PSA-GM alliance was established during a difficult economic and social period. In 2012, the European car market was shrinking considerably (−6 % sales), PSA saw its sales drop by 7.7 % and GM by 15.6 % during the same year. Both manufacturers experienced operating difficulties following this decrease in sales on the European market and had to deal with a decline in profitability. Overcapacity problems for both carmakers put pressure on their alliance, such as at the Aulnay plant in France for PSA and Bochum in Germany for Opel (GM subsidiary) (Jullien 2013).

In addition, the recession in Europe of around 0.4–1 % had a direct impact on private demand, which further reduced car purchases and increased decisions to postpone purchase. In a structural manner, new modes of mobility emerged and created new vehicle usages for consumers, such as carpooling and car sharing, which have continued to grow. The progression of alternative means of transport and restricted vehicle access in major towns also had an impact on the already very fragile European car market.

In the above context, how could a large-scale alliance to increase volumes be established? Too many factors point to weaknesses in the joint construction of this type of alliance. The first is the consideration of a recessive economic environment that made it necessary for both carmakers to globally rethink their commercial strategy; the second concerns a financially fragile situation, which was certainly different for PSA and GM, but which led to defensive and restrictive choices. PSA thought it would improve its financial flexibility by raising 1 billion euro in 2012, of which 300 million represented the 7 % of PSA shares paid by GM. Yet, in 2013, PSA announced an emergency financial rescue plan to deal with a 4.5 billion euro deficit, which had the effect of the Chinese group entering PSA's capital at 30 % (PSA Peugeot Citroën 2014).

Lastly, the constraints of this alliance, far from creating development opportunities, added to earning losses, since PSA was forced to abandon markets. Thus, Iran, PSA's second market behind France, was abandoned despite annual sales of 472,000 vehicles in 2010. Similarly, PSA saw its sales drop significantly in Brazil, where it was incapable of matching its formidable competitor: its ally, GM!

This brief history raises numerous questions. How should we interpret this partnership failure? Why did the carmaker attempt new rapprochements so rapidly after the demise of a partnership? Did this new strategy pay off?

---

[5]Journal *La Tribune*, 14 Décembre 2013.

5.3.2.2 Faced with New Strategic Challenges, an Innovative Partnership Between PSA-Peugeot-Citroën and IBM

The partnerships that we have described between carmakers are already part of automobile history. Today, automobile manufacturers have to face new challenges and a new ecosystem in the making with the connected car. Numerous questions are being raised in the car industry, such as: How to manage a new cooperation set-up with players that have not up to now been involved with their trade? How to generate added value in constantly evolving production modes?

The Business Unit Smart Car was set up in September 2013 by PSA Peugeot Citroën to respond to this challenge. The carmaker proved its capacity for innovation in connected services when it created the emergency call ten years ago, and decided to capitalize on this knowhow to develop a new PSS (products/services/systems) approach.

When asked about the objective of this Business Unit, strategy manager Y. Bonnefont explained that it involves, "*Developing services based on connected and digital cars that will make it possible to respond to new uses and requirements in terms of mobility, and in particular, create an experience that strengthens customer loyalty.*" This point of view is widely shared by PSA's Connected Services and Mobility Business Unit Director, B. Courtehoux, who steered the PSA Peugeot Citroën/IBM alliance. These two partners got together to tackle a new phase in the launch of "connected services" for tomorrow's car. The aim is to accelerate the development of services tailored to customers. "*We want to be capable tomorrow of capturing data from our connected cars to offer drivers innovative, targeted services.*"

Thanks to Big Data and Analytics and IBM's MobileFirst,[6] customer relationships are completely changing and opening the way for a totally transformed mobility framework. For B. Courtehoux, "*Analyzing the collected data will give us precise knowledge of vehicles' faults and drivers' behavior. We will ascertain which functions customers use the most, those that they never use, and for example, how many times a year they open the panoramic roof of their car. It will allow us to optimize vehicle design and adjust the price*" (L'usine digitale 2014). Every connected car potentially features several thousand data collected from around a hundred embedded sensors. One example among many is the meteorological precision resulting from embedded temperature, anti-fog and windscreen-wiper sensors.

In this area, the question arises of capitalizing on the knowledge of all employees, along with their agility and/or adaptability to integrate this mass of knowledge into eco-design innovations. For PSA's Connected Services and Mobility Business Unit Director, "*The real knowhow lies in human intelligence and in our data scientists' capacity to model the application and identify what information to extract from the huge volume of data*" (idem). The jobs-trades concerned (development, production, quality) are stakeholders in the project, associated with other trades and services like IT.

[6] IBM website, www.ibmbigdatahub.com.

Management is a constant challenge for the Direction and the teams in this context. Indeed, it involves the interweaving of highly diverse cultures-trades, from IT to design, and production to connected services. Another difficulty resides in managing knowledge, with the constant integration of gigantic, diverse volumes of collected data. It is a major challenge for the carmaker to integrate the added value brought from the processing and *real-time analysis* of Big Data coming from connected vehicles.

Implementing a new management method necessarily involves new forms of governance. The success of these alliances thus depends on the existence of a "moral community", relationships of trust, and mutual understanding that ties in with the concept of *embeddedness* developed by Granovetter in 1985 and taken up by Donada et al. (2012) and Donada and Attias (2015) to characterize the governance modes of these new partnerships. Indeed, in the electro-mobility industry, embedded governance could be a better way to identify the difficulties of sharing value between players that have neither the same trades nor the same production purposes (Donada and Attias 2015).

5.3.3 The Autonomous Car at the Heart of New Alliances?

In the current mobility ecosystem, connected and/or autonomous cars are drastically changing the positions of traditional players, and several scenarios seem possible, either in terms of alliances between carmakers and complementary companies, or through the emergence of new players from *outside* the automobile sphere.

We have already mentioned a radically new type of alliance in the history of the automotive industry in France, between carmakers and IT service companies and/or communications companies (cf. PSA Peugeot-Citroën and IBM). This type of partnership is becoming widespread: the Spanish car manufacturer Seat started with an alliance with the Korean telecommunications and electronics company Samsung to add a connection interface to its cars with smartphones, and then extended this alliance to SAP, the world leader in enterprise applications and business networks. Together, the three companies are targeting the future market for connected cars. The project unveiled in 2016 involves developing a virtual key to identify vehicle users and a connection to access and pay for parking places from a telephone. The challenges are considerable in this type of partnership, which puts into question traditional carmakers, which will most likely tomorrow be car service providers and no longer car manufacturers.

Several reasons are involved in this scenario: the creation of value generated by service contracts (battery hire, availability of autonomous driving software, memory card to manage traffic or weather, etc.) will be greater than that initially created by the manufacture of cars. In all industrialized economies, we observe the displacement of added value from products to services (Gadrey 2003). The automotive industry is no exception to this economic trend, which has considerable implications: deindustrialization, a drop in profits, and vehicle manufacture moved to emerging countries. Most of the income resulting from these service sales already goes to companies other than the carmakers.

In addition, new alliances are emerging in these mobility service companies, raising the question of tomorrow's leadership. The TomTom and Bosch alliance forged in 2016 is a case in point. These two companies decided to work together on high-precision mapping, which is essential to develop automated driving. It involves an alliance between two automobile service suppliers that want to take advantage of the opportunities offered by the development of a connected vehicle without working closely with any carmakers for the time being. TomTom will bring its knowledge of mapping and traffic monitoring, while Bosch will define the mapping precision required based on its systems management experience.

In China, in March 2016, the announcement of an alliance between the Chinese automobile manufacturer SAIC and the Chinese electronic commerce giant Alibaba created quite a stir (L'usine digitale 2015). A billion-yuan investment to create a fund dedicated to developing connected cars is, according to the two companies, the first stage in building alliances with other investors of new technologies and services using cloud computing.

These companies, some of which are far removed from the automobile sector, have understood the economic and financial issues of a new mobility ecosystem and the potential for capturing the value created. This involves a genuine reconfiguration of the automotive industry and services. However, these new mobility players face competition themselves from start-ups like Lyft, identified by General Motors and already competing with Uber, which is trying to position itself against driverless cars and taxi robots, like those presented in 2015 by a Japanese company that has announced the first trial runs south of Tokyo in 2016 (CNET 2015)

Other entrants with high financial capacity, like the carmaker Tesla, which is also developing a new business model, are stakeholders in this new mobility ecosystem and their market share continues to grow. According to the financial services firm Morgan Stanley, Tesla Mobility 1.0 is a new business model, known as Position on Demand Service (PODS). Tesla's new car will be electric, autonomous and shared. "*100 % of Tesla's cars are electric, connected, and able to 'learn' through over-the-air firmware updates at any time*" (Jonas 2015). Will Tesla be the best placed to lead the market? The global mobility area is transforming radically with fast technological changes, meaning that traditional carmakers all have an obligation to apprehend this new reality in a global way, integrating new requirements, new services and new uses.

5.4 Conclusion

From business ecosystems to innovation ecosystems, the world of the automobile was built up over time, with alliances, collaborations and partnerships whose purpose was to pool technical and financial resources while retaining a dominant share of the market. The shift to the mobility ecosystem has created a new innovation momentum, constrained carmakers to think up a new economic model, and, with the arrival of new entrants, led to a reorganization of the entire automobile sphere.

In this new mobility ecosystem, "traditional" forms of coopetition in the automotive industry are being rethought because the automotive industry is reinventing itself. Economic models are radically changing, new alliance strategies that cut across trades and sectors are now the keystone of the car industry's development. These new economic models are opening up a debate on the creation and sharing of value. Automobile manufacturers, constrained to be constantly agile, must reinvent their strategy for a new century of development led by innovation and aiming at sustainable, responsible mobility.

Acknowledgment This research benefits from the support of the Chair "PSA Peugeot Citroen Automobile: Hybrid technologies and Economy of Electromobility", so-called Armand Peugeot Chair led by CentraleSupélec, ESSEC and sponsored by PEUGEOT CITROEN Automobile.

Bibliography

Barnes, S.J., 2002. The mobile commerce value chain: analysis and future developments. International journal of information management 22, 91–108.

Bengtsson, M., Kock, S., 2000. "Coopetition" in business Networks—to cooperate and compete simultaneously. Industrial marketing management 29, 411–426.

Brandenburger, A., Nalebuff, B., 1996. Co-opetition (New York: Currency Doubleday).

Cazenave, F., 2014. Stop Google: Relever les nouveaux défis du géant du web. Pearson Education France.

Chesbrough, H.W., 2006. Open innovation: The new imperative for creating and profiting from technology. Harvard Business Press.

Christensen, T.B., Wells, P., Cipcigan, L., 2012. Can innovative business models overcome resistance to electric vehicles? Better Place and battery electric cars in Denmark. Energy Policy 48, 498–505.

CNET, 2015. Les taxis robots arrivent au Japon dès 2016 [WWW Document]. CNET France. URL http://www.cnetfrance.fr/cartech/les-taxi-robot-arrivent-au-japon-des-2016-39826440.htm (accessed 3.11.16).

Dammenhain, K., Ulmer, T., 2012. Managing the change to emobility. Capgemini.

Di Minin, A., Frattini, F., Piccaluga, A., 2010. Fiat: open innovation in a downturn (1993–2003). California Management Review 52, 132–159.

Donada, C., Attias, D., 2015. Food for thought: which organisation and ecosystem governance to boost radical innovation in the electromobility 2.0 industry? International Journal of Automotive Technology and Management 15, 105–125.

Donada, C., Attias, D., 2013. Innovative Project Management in the Automotive Industry. Journal of Modern Project Management 44–49.

Donada, C., Fournier, G., 2014. Stratégie industrielle pour un écosystème en émergence: le cas de la mobilité 2.0, décarbonée, intermodale et collaborative. Revue d'économie industrielle 317–348.

Donada, C., Nogatchewsky, G., Nogatchewsky, S., 2012. Gouvernance interorganisationnelle imbriquée et stratégie orientée client. Comptabilité-Contrôle-Audit 18, 7–32.

Fernandez, A.-S., Le Roy, F., 2013. Comment coopérer avec ses concurrents? Revue française de gestion 61–80.

Fréry, F., 1997. La chaîne et le réseau. Dedans-dehors, les nouvelles frontières de l'organisation 23–53.

Gadrey, J., 2003. Socio-économie des services, 3rd edition. ed. La découverte Paris.

Ghosn, C., Riès, P., 2005. Citoyen du monde. Librairie générale française.

Iansiti, M., Levien, R., 2004. The keystone advantage: what the new dynamics of business ecosystems mean for strategy, innovation, and sustainability. Harvard Business Press.

Ili, S., Albers, A., Miller, S., 2010. Open innovation in the automotive industry. R&d Management 40, 246–255.

Isckia, T., 2010. Ecosystèmes d'affaires, stratégies de plateformes et open innovation, 19ème conférence de l'AIMS, Luxembourg.

Iyer, B., Davenport, T.H., 2008. Reverse Engineering: Google's Innovation Machine. Harvard Business Review 86.

Jonas, A., 2015. Top Wall Street analyst dreams up a whole new business model for Tesla and predicts the stock goes up 91 % [WWW Document]. Business Insider. URL http://uk.businessinsider.com/morgan-stanley-raises-tesla-price-target-2015-8 (accessed 3.12.16).

Julien, P., Raymond, L., Jacob, R., 1996. L'Entreprise-Réseau: Dix Ans d'Expérience de la Chaire Bombardier, Produits récréatifs, Presse de l'Université du Québec.

Jullien, B., 2013. Gerpisa Seminar, Ecole Normale Supérieure de Cachan, 17 December 2013.

Jullien, B., 2000. La distribution automobile en 2000: vers le partenariat ou la taylorisation. Presented at the 8th GERPISA international Colloquium The World that Changed the Machine: The Future of the Auto Industry for the 21st Century. Paris, June, pp. 8–10.

KPMG, 2010. KPMG's Global Automotive Executive Survey 2010. Industry Concerns and Expectations to 2014.

Le Roy, F., Yami, S., Dagnino, G., 2010. La coopétition: une stratégie pour le vingt-et-unième siècle. Les stratégies de coopétition: rivaliser et coopérer simultanément, DeBoeck, Collection Méthodes et Recherches, Bruxelles 17–28.

Livian, Y.F., 1998. Introduction à l'analyse des organisations, Paris, Economica 1995, 2nd ed. 1998.

Loubet, J., 2004. Histoire de l'automobile française, Paris, Le Seuil.

L'usine digitale, 2015. Alibaba s'allie avec le constructeur chinois SAIC pour travailler sur la voiture connectée [WWW Document]. usine-digitale.fr. URL http://www.usine-digitale.fr/article/alibaba-s-allie-avec-le-constructeur-chinois-saic-pour-travailler-sur-la-voiture-connectee.N318818 (accessed 3.11.16).

L'usine digitale, 2014. PSA veut valoriser le Big Data de ses voitures connectées [WWW Document]. usine-digitale.fr. URL http://www.usine-digitale.fr/article/psa-veut-valoriser-le-big-data-de-ses-voitures-connectees.N250912 (accessed 3.11.16).

Magee, D., 2005. Comment Carlos Ghosn a sauvé Nissan. Dunod.

Mahut, F., Daaboul, J., Bricogne, M., Eynard, B., 2015. Survey on product-service system applications in the automotive industry. IFAC-PapersOnLine 48, 840–847.

Métais, E., Pin, R., 2002. De la subversion à la normalisation: de la Swatchmobile à la Smart. Gestion 27, 14–21.

Miller, R., Olleros, X., Molinie, L., 2008. Innovation games: a new approach to the competitive challenge. Long Range Planning 41, 378–394.

Moore, J.F., 1996. The death of competition: leadership and strategy in the age of business ecosystems. HarperCollins Publishers.

Moore, J.F., 1993. Predators and prey: a new ecology of competition. Harvard business review 71, 75–83.

Nieuwenhuis, P., Wells, P., 2003. The automotive industry and the environment: A technical, business and social future. Cambridge, Woodhouse Press.

Park, S.H., Russo, M.V., 1996. When competition eclipses cooperation: An event history analysis of joint venture failure. Management science 42, 875–890.

PSA Peugeot Citroën, 2014. PSA Peugeot Citroën Sustainable Development and Annual Report 2013.

Ramirez-Portilla, A., Brown, T., Cagno, E., 2014. Open Innovation in the Automotive Industry: What can carmakers' annual reports tell us? Presented at the The 8th International Conference on Industrial Engineering and Industrial Management-CIO-ICIEOM-IIIE, Malaga (Spain), 23rd–25th July 2014.

Rishi, S., Stanley, B., Gyimesi, K., 2008. Automotive 2020: Beyond the chaos. Somers, NY: IBM Institute for Business Value.

Rogers, E.M., 2003. Diffusion of Innovations, 5th edition. ed. Free Press, New York.

Rugman, A., D'Cruz, J., 2000. Multinationals as Flag-ship Firms: A New Theory of Regional Business Network, Oxford University Press: Oxford.

Sarasini, S., 2014. Electrifying the automotive industry: The geography and governance of R&D collaboration. Environmental Innovation and Societal Transitions 13, 109–128.

Thorelli, H.B., 1986. Networks: between markets and hierarchies. Strategic management journal 7, 37–51.

Wells, P., 2013. Sustainable business models and the automotive industry: A commentary. IIMB Management Review 25, 228–239.

Wells, P., Nieuwenhuis, P., 2012. Transition failure: Understanding continuity in the automotive industry. Technological Forecasting and Social Change 79, 1681–1692.

Word, J., 2009. Business network transformation: strategies to reconfigure your business relationships for competitive advantage. John Wiley & Sons.

Chapter 6
Smart Cities and Smart Mobilities

Patrice Geoffron

Abstract Sociodemographics are transforming the world into an 'archipelago of cities', with over 70 % of the world's population concentrated in urban environments by 2050. This trend brings as many threats (impact of pollution on public health, economic losses caused by congestion) as perspectives for new urban organizations. In this context, 'smart cities' are emerging. Although heterogeneous, smart cities have in common the optimization of data management to improve urban services, i.e. transport, energy, waste, habitat, health, education and culture. The issue of transportation crosses over all aspects of smart cities, whether in terms of urban design and social organization (more compact towns and distance work to reduce flows), or organizing new ways to manage vehicle capacities and infrastructure (shared fleets, car sharing, urban charging, road lane management), combined with the mid- or long-term dissemination of incremental innovations (electric vehicles) or disruptive innovations (autonomous vehicles). For the traditional automobile ecosystem (car and equipment manufacturers, etc.), the emergence of smart cities constitutes a potentially disruptive challenge with the calling into question of combustion cars in towns, new competition with other industrial players (information technology, community services, utilities, etc.) and the diversification of economic models (reliance on big data, less ownership, more service-rich).

Keywords Smart city · Electric vehicle · Energy transition

6.1 Introduction

The world's metropolises reinvented themselves at the start of the 21st century, echoing the urban planning of the late 19th century with the development of electricity, transport and water networks. This time round, new urban models are emerging in response to a technical 'impulse' (the dissemination of information and communica-

P. Geoffron (✉)
Paris-Dauphine University, Pl. du Maréchal de Lattre de Tassigny, 75016 Paris, France
e-mail: Patrice.Geoffron@dauphine.fr

D. Attias (ed.), *The Automobile Revolution*, DOI 10.1007/978-3-319-45838-0_6

tion technologies) and to deal with the sociodemographic pressure transforming the world into an 'archipelago of cities': since 2006, more than one in two people live in a town, a proportion that could reach 70 % by 2050, and over 80 % in OECD countries, for a global population of around 10 billion people by then. This urbanization of the world has another side: urban concentrations already produce 80 % of the world's GDP, so that the capacity to organize the convergence towards cities will depend on economic prosperity and, without doubt, geopolitical stability.

The urban transformations at the start of this century are focused on city projects known as 'smart cities' that are in reality highly heterogeneous (especially depending on whether they are in rich or emerging countries). What these smart city projects do have in common, however, is the optimization of data management to improve urban services, i.e. transport, energy, waste, health, education and culture.

One of the crucial drivers of these transformations is to reduce urban pollution and anticipate the risks of 'thrombosis' that automatically result from this continued exodus towards 21st-century cities. The development of 'smart mobility' is therefore key to the transformations that lie ahead. This smart mobility issue reveals a profusion of solutions that cities will combine into specific local models, such as transformations of urban areas with more compact housing projects, dynamic traffic management, extended multi-modalities, etc. Naturally, the presence of combustion vehicles in towns is questionable, with the more or less long-term dissemination of substitutes for motorization (electric cars), piloting (autonomous vehicles), ownership modes (car sharing, car pooling), and economic models (more service-rich), etc. Since this transformation involves players that have until now been peripheral to the automobile industry (GAFA, utilities, community service groups), a new competitive area is being drawn up, with data management as the focal point.

To analyze the context and development forms of this smart mobility, we work in three stages: first we present urban transport issues in the context of the low-carbon energy transition (Sect. 6.2); then we define the process whereby smart cities emerge (by suggesting a typology to delimit their diversity) (Sect. 6.3); and lastly we analyze the issue of mobility, which is also "smart", in these new urban models (Sect. 6.4).

6.2 The Challenges of "Low Carbon" Transport in the Global Urbanization Process

Although in 2015, COP 21 culminated in the Paris Agreement subject to ratification by almost 200 countries, local stakeholders were particularly active in the run-up and during negotiations: the NAZCA (Non state Actor Zone for Climate Action) platform listed over 7000 local authorities, 2255 towns and 150 regions, representing almost 20 % of the global population, that made climate commitments in Paris. Many of these commitments relate to the transport field; thus the initiative *MobiliseYourCity* aims to support 100 towns to draw up sustainable urban mobility plans. Each participating town commits to a 50–75 % reduction in emissions from its urban transport system by 2050.

The proactive approach of these pioneer towns (generally more ambitious than the countries they are located in) reflects the complexity of climate issues, which are both "global" (reduction of greenhouse gas emissions) and "local", with the need to combat the *airpocalypse* episodes (periods of intense urban smog) that are recurrent in some metropolises, and anticipate the effects of climate change (by developing resilient urban infrastructures). The annual report from the *World Health Organisation* (2016), covering 3000 cities in over 100 countries,[1] calls for urgent action: since 2010, global levels of atmospheric pollution in urban environments have gone up by 8 % and cause almost 4 million premature deaths every year. The origins of this deterioration are well known: for the most part the rapid expansion of motorized transport combined with dependence on fossil fuels in electric power plants.

This urgency is encouraging some towns, in an extension of the COP, to undertake proactive, even disruptive action. In India, New Delhi is obliging its taxis to start using natural gas (which creates very low local polluting emissions and generates less CO_2 than petrol). In Europe, about twenty major cities (including Paris, Copenhagen, Milan, Amsterdam and Stockholm) have, in opposition to the European Commission, embarked on a legal battle before the Court of Justice of the European Union regarding the decision to reduce thresholds for polluting car emissions.

These strong actions by cities are a response to the current threat of urban congestion and pollution and, in particular, the perspective of their endemic character in the face of the structural developments ahead. In 2015, 52 % of the world's population lived in towns (whose surface area only covered 2 % of the globe), a proportion likely to reach 70 % by 2050. Urban areas also dominate primary energy consumption (65 %) and greenhouse gas emissions (70 %). If the trends observed at the start of the 2010s continue, energy consumption will grow by another 70 % and emissions by 50 % by 2050 (IEA 2016).

In this perspective, transport contributes around 25 % to greenhouse gas emissions and, because incremental requirements are rising sharply in the emerging and developing world, a 20 % increase is likely by 2030 and a 50 % rise by 2050. Since half of passenger transportation currently takes place in urban environments (representing 40 % of the energy used in terms of transport), the world's transformation into an 'archipelago of cities' by the middle of the century calls for a drastic turnaround in our current models.

To develop urban transportation in line with the COP 21 targets does not necessarily constitute an economic challenge, since it could lead to the massive

---

[1]Although most data come from cities with populations of over 50,000, one quarter covers zones that count less than 20,000 inhabitants, so that this panorama also reflects the level of air pollution in small built-up areas and not just giant metropolises. Similarly, it should not be considered that the costs of pollution are only significant in emerging countries. According to the OECD and the WHO (WHO-OECD 2015), the deterioration in air quality in Europe amounts to a cost of 1400 billion euro per year, which represents from 1 to 10 % of the gross domestic product (GDP) of countries covered by the study: 2.3 % for France, 3.7 % for the United Kingdom and 4.5 % for Germany.

reduction of investments in infrastructure and vehicles in comparison to today's models.[2] Despite this potential for economic rationalization, the challenges are numerous in terms of reorganizing complex systems. Thus, the wide dissemination of the electric vehicle is not only based on improving autonomy, but on reorganizing electricity supply networks (using smart grids, in other words grids that can interconnect the elements of the system via a sophisticated telecommunications network) to respond to additional requirements, as well as decarbonate production (with the risk of having vehicles in reality recharged with coal). More fundamentally, it will not be possible to adapt transport systems to the challenges of endemic congestion and pollution without changing cities themselves, either moving towards more compactness or reorganizing human activities within them to rationalize internal flows. This is the background for the debate on transforming metropolises into smart cities.

6.3 Smart Cities: New Model or New Utopia?

The management of urban transport flows is part of a much larger issue in that, in the context described above, aims to reorganize the infrastructures that make up towns, enrich the services delivered to their inhabitants and, beyond that, involve those inhabitants in their co-production. This archetypal smart city is a continuation of projects started at the turn of the 20th century, when urban planning started to take its modern form, and the era of mechanization and industrial development began.

The result was a utopian city that would be both functional and purged from pollution (caused by coal in 19th century London, just as it is in 21st century Beijing). At the start of the last century, the panorama also included a wealth of new technologies that influenced the organization and representations of new urban forms (hydroelectricity, automobiles, aviation, cinema, photography, etc.), artistic movements (such as Italian Futurism in the 1910s) that presented a triumphant vision of technology, and architectural movements (Bauhaus in Germany in the 1920s) that idealized an architecture adapted to the "world of machines" (Hall 2002). The architect T. Zenetos, prefiguring the supply of smart cities via networks, even designed a model of "Electronic Urbanism", which was highly visionary in the way it envisaged telecommunications (especially distance work and education) (Zenetos 1969).

The term smart city emerged in the early 1990s in the work of Batty (1990) and Laterasse (1992), but did not become the focal point of urban planning until around 2010, although its definition was still not stable. What is unusual about the smart city in comparison with the design of modern towns that came before it, is the original combination of technology push and demand pull (Angelidou 2015).

[2]According to the International Energy Agency (IEA 2016), opting for the COP 21 2 °C scenario, compared to business as usual (likely to lead to a 6 °C increase by the end of the century), would reduce the need for investments in urban transport systems by USD 21 trillion, partly due to the reduced personal vehicle stock.

On the supply side, smart cities are initially built up around a clutch of innovations in the information technology domain, involving the deployment of fixed and mobile electronic networks converging at the "internet of things" (in other words the connection of all public and household equipment to the Internet), the uses made of them (horizontal interconnections through social networks), and the multiplication of screens and new interfaces (from smartphones to intelligent windscreens) suitable for augmented reality experiences. Because these intensely connected and innervated cities contain a profusion of data, the conditions of their dissemination (whether open data or not), the capacity for processing them, and their exploitation, make way for new innovations that are likely to lead to disruption (sharing transport or housing capacities between individuals) when "two-sided" platforms (Airbnb and Uber for example) establish themselves as reference intermediaries.[3] Although the disruptive potential of smart cities results from electronics (and the phenomenon of interconnection platforms), it is the combination of technical progress in buildings and energy, making it possible to design positive energy buildings (associating thermal efficiency, insertion of renewable energies, energy storage capacity) and connect them together in ecodistricts, that makes it possible to draw up new urban landscapes (without mentioning those naturally organized in the transport domain that are the subject of the next section).

On the demand side, in addition to the challenges of human convergence in towns and the management of pollution and congestion, the financial crisis of the late 2000s increased pressure on public authorities to improve management of public resources while maintaining (or promoting) high service levels and diversifying them. In the context of globalization, metropolises compete to attract investments, highly qualified human capital, tourists, international events, etc. This is clearly the approach, for example, taken by the European towns with the most proactive smart strategies, such as Amsterdam, Barcelona and Lyon. However, this kind of specification leads to complex ecosystems of social groups and institutions and naturally raises questions about how to organize democracy and make 'society', with contradictions between inclusive rationales (joint construction of services by inhabitants, participative democracy) and exclusive ones (threat of gentrification) (Bélissent 2011; Ratti and Townsend 2011).

Note, in addition, that the distinction between service suppliers and consumers is less clear in a smart city context. Applying information and communication technologies to towns calls for increasing investment from users, who need to be more closely involved with managing the new services offered to them. This is the case of

[3]Two-sided platforms interconnect categories of economic agents with interdependent interests. This means that the benefit of an agent located on one 'side' is connected to the number of agents located on the other (or the breakdown of that group), a phenomenon known as 'crossed externalities'. This configuration results in very specific competitive tensions. The indirect network effects constitute a concentration factor ('snowball' effect): an increase in the number of agents on one side is likely to attract more to the other side and vice versa in a rising spiral with the possibility of the emergence of a monopolistic platform in fine (known as 'winner takes all') (Rochet and Tirole 2003).

smart grids and new electricity (and gas) production and supply networks, since users are likely to act on the energy supply (including via production capacities integrated into their homes), modulate their demand (depending on production constraints and especially the CO_2 emitted from the electricity produced, by reacting to price signals), and even to contribute to the balance of the system via their electric vehicles' storage capacity ('vehicle to grid', 'vehicle to home').

Although the configurations of smart cities, beyond the general orientations mentioned above, are characterized by their heterogeneity (geographic, economic and sociological conditions culminating in local "idiosyncratic" needs), it is possible to distinguish dominant models according to specific levers of action (Ecube 2010):

- 'Eco-Cities' are towns built with an ecological approach that emphasizes sustainable development and reduces the carbon footprint, such as Stockholm. The levers of action are the choice of building materials, bioclimatic architecture, urban design engineering, etc. These projects might involve new land planning zones in a small areas (e.g. BedZED, covering 2 hectares in south London[4]) or new towns (Dongtan, north of Shanghai, covering 90 km^2 and aiming to house 500,000 residents by 2050). The economy of these 'Eco-Cities' is based on highlighting externalities (avoidance of CO_2 emissions, local pollution and congestion).
- 'Smart Grid Cities' are organized around the active management of networks and electricity consumption to make it easier to integrate renewable energy or electric vehicles. A pioneer experiment in this area was Boulder (Colorado) in 2008, under the impetus of the American electrician Xcel Energy. The economic model is based on avoiding carbon energy production costs and related network costs, and the use of electric means of transport.
- 'Ubiquitous Cities' (or 'U-Cities') are run by operators and telecom equipment suppliers based on the idea of pooling communication infrastructures and sharing information. One of the first experiments is the Hwaseong-Dongtan Connected City (in 2007) in South Korea, proposing new services connected to transport and security.

6.4 Smart Mobilities: Between New Designs and More Electronic Cities

Concerning the issue of transport in smart cities, it is important to bear in mind that the extent of urban spread is crucial in determining systems' sustainability and efficiency. Urban compactness is one of the main conditions when deploying sustainable systems, since it helps reduce journey distances and allows investment in high-capacity public transport, while facilitating journeys by foot and bicycle (OECD 2010). Although it is difficult to make very sprawling cities more compact, it is possible to support the growth of emerging metropolises by following a

[4]Beddington Zero Energy Development.

polycentric model, with subdivisions that combine habitat and tertiary activities (or even industrial ones) and so delimit journeys, like the expansion model adopted by Shanghai in the 2000s (Henriot 2016).[5]

Due to their density, and in line with the European Union's climate and environment targets, European towns are among the most advanced in terms of experimenting with new mobilities, through defining new regulations (incentive tariffs in Stockholm, Milan, etc.[6]), dividing up urban areas ("low emission" zones in Stuttgart, Berlin and Lisbon, etc.[7]), redeploying economic activities (co-working areas spread around the city of Amsterdam), disseminating noise-free modes of urban transport (bicycles and electric vehicles in Paris and Lyon), and interconnecting towns with "soft" transport (the Ruhr is served by a 100-km highway for non-motorized bikes).

This diversity confirms that the transportation issue concerns all smart cities and is one of its most complex dimensions, in as much as the solutions applied have a strong impact on the way public areas are organized.[8] In particular, it involves managing basic contradictions between the need for mobility (resulting in particular from economic activity) and the attractiveness of a town interwoven only by transportation producing little pollution (in terms of gas and particle emissions and noise disturbances). The example of urban deliveries perfectly illustrates this kind of contradiction: managing the 'last mile' is a particularly costly activity and a potential source of highly negative externalities (in particular in historical town centers); although urban deliveries only represent 1 % of freight transport (in km covered), they consume 21 % of the energy associated with this type of activity.[9] Yet the development of electronic commerce increases the need for city-center

---

[5]Shanghai has adopted a polycentric expansion approach with the deployment of new towns in its 'One City, Nine Towns' project launched in 2001. More generally, over 200 smart city projects are under experimentation in Chine.

[6]Although road charging and congestion charges are rarely employed, analyses show that they have a significant impact on modifying uses and reducing externalities (Azari et al. 2013), which goes to explain their integration in scenarios corresponding to COP 21 targets (IEA 2016).

[7]Initially developed in Europe, geographical restrictions (e.g. pedestrian zones, prohibition of types of polluting vehicles at some times of day or during peak hours) are rapidly being developed in Asia, especially China.

[8]Whereas other attributes of smart cities (environmental qualities of the habitat, introduction of green roofs, access conditions for some remote public services, etc.) fit in more easily with pre-existing urban constraints.

[9]The intensity of greenhouse gas emissions from urban deliveries is 3.5 times higher than for long-distance transport per ton-kilometer. The case of Barcelona illustrates both the constraints and solutions in a smart city approach: "Goods vehicles make up 6.6 % of Barcelona's vehicle stock, but they constitute only 15 % of city traffic and 23 % of connecting trips. The city council of Barcelona has recently approved the Sustainable Urban Mobility Plan (SUMP) 2013–2018, which focuses specifically on logistics activities. In particular, [...] inside the optimisation of urban goods, it specifies a way: 'Establish a network of multi-use vehicle parks and mini trans-shipment platforms in neighbourhoods from which goods deliveries can be realised by trolleys, electric vehicles and tricycles/cargo-bikes'" (Navarro et al. 2016, p. 316).

deliveries, since some of the competition between major players depends on their quality of service in this area.[10]

Just as smart cities do not come under one single definition, the notion of smart mobility is subject to a set of very different solutions. These do, however, differ from 20th-century experiments in their use of digital technology to optimize the multimodal chain, and the nature of value-added services offered to users. This digitalization of mobilities, whether in place or in perspective, can be characterized as follows.

In a great number of metropolises, car traffic is initially monitored in real time using techniques ranging from induction loops that identify the passing of vehicles, video cameras, widespread use of sensors, and analysis of related data. This information-rich environment means that users can adapt their behavior, whether in terms of choosing a multimodal strategy, the conditions for using roads with a personal vehicle, or car parks (on public roads and in private spaces). For public authorities (or private companies under public contract), digitalization also results in more dynamic management, allowing them to adapt tolls in real time (according to observed constraints) or even modulate the allocation of the road network (by prioritizing public transport or pooled vehicles). One of the challenges is to avoid restricting dynamic management to organizing flows, and instead cross these data with data from sensors monitoring environmental quality, public lighting, cleanliness, etc. The objective is to organize information platforms (with an open data approach) so that administrations and/or private companies can put together innovating services, like Masdar's pioneering experiences in Abu Dhabi or Songdo's in South Korea.[11]

Since the digitalization of towns goes hand in hand with that of its inhabitants, the latter's interface equipment (especially smartphones) means they can access both traffic information in real time (for car circulation), and data relating to the state of public transport (times, frequency, incidents, etc.). It also means that shared urban fleets can be deployed, such as bicycles, self-service electric cars and scooters, by geolocalizing pools of vehicles, and return parking areas. When inhabitants possess a significant level of equipment, car-sharing platforms can be organized (e.g. Blablacar, and intra- or inter-enterprise systems) or chauffeur-driven car services using the fleet of privately owned vehicles or employing private car owners as drivers (e.g. Uber). All of these services, whether they are promoted by municipal authorities or developed by third-party companies organizing platforms, introduce a fundamental change in that they reduce the need for directly possessing a vehicle (motorized or not). The types of coordination that information technologies make possible thus generate a wide diversity of models oriented towards sharing, membership or subscription to services, rather than acquisition. This

---

[10]Inaugurated in New York in 2014 and now available in almost 50 cities around the world, Amazon's "Prime Now" system offers express home delivery within an hour (the average time is 36 min) for all types of goods, including food, 7 days a week from 8 am to 10 pm. This "innovation" illustrates the pressure, particularly from customers with high purchasing power, on service quality, leading to competition in the delivery conditions sector.

[11]This is the case of the "connected boulevard" experiment carried out by the city of Nice in association with Cisco Systems.

phenomenon fits in with the "smart mobility" movement in as much as it is a result of information technologies. However, transportation's environmental impact on cities differs depending on whether the innovations introduce soft multimodality (with in particular the development of journeys by bicycle), or whether they create more flexibility in the use of automotive vehicles, with uncertain global impacts (depending on whether these uses replace the ownership of a private vehicle or the use of public transport).

In this environment, electric vehicles fulfill the conditions of higher acceptability in city centers (whether for transporting people or making deliveries), while obeying specific constraints because of their dual interconnection to the electricity grid and a telecommunications network. Although the global stock of electric vehicles is still very limited in the middle of the 2010s (around one million), attaining the COP 21 targets will involve the widespread use of this type of motorization: the global stock of electric cars and freight delivery vehicles will need to exceed 150 million by 2030 (9 % of the total fleet) and approach 1 billion by 2050 (about 40 %) (IEA 2016).

However, beyond the advantages of the motorization of electric vehicles compared to internal combustion engines,[12] their expansion in smart cities involves a systemic dimension within electric smart grids, the constraints and potential of which need to be anticipated:

- The sizing of public supply networks and the balance between electricity production and consumption is impacted by the deployment of electric vehicles, with the risk of a significant increase in peak consumption. The choice of when to recharge and at what power level needs to take into account the constraints on the grid. It is therefore important to raise user awareness of these network constraints and of the variable carbon content of electricity depending on the period. This implies price signals, as well as automated control of recharging.
- Embedded batteries in electric vehicles can also be used to store energy. Since any given vehicle is stationary most of the time, the grid can draw from the battery the electricity required to satisfy consumption peaks or compensate a drop in photovoltaic or wind power production. With this approach, known as vehicle-to-grid (V2G), vehicles bring flexibility to the electric grid that fuels them.[13] Batteries can become a source of energy to supply households, in an approach pioneered by Japanese constructors and known as vehicle-to-home (V2H).

[12]On the condition, however, that the production system is low-carbon. For example, although electric vehicles circulating in Chinese towns have the advantage of distancing some types of pollution from city centers, their deployment does not fundamentally reduce greenhouse gas emissions. Other types of motorization, for example natural gas vehicles, present fewer constraints (considering the necessary adaptation of the electricity grid), while presenting an acceptable environmental footprint (contained CO_2 emissions, near absence of local pollutants).

[13]Since the considerable electricity crises of the early 2000s, California has been looking closely at solutions for securing the local electricity supply. The state's transport operator, California Independent System Operator (CAISO), has developed an experimental V2G system as part of its strategy.

The development of autonomous vehicles (which are already authorized to circulate in some US states), both electric and combustion, constitutes the next stage in the move towards smart mobilities. This kind of project, which most major automotive groups are currently working on, brings great benefits within the perimeter of smart cities, in terms of accidents, congestion management, vehicle compactness (whose security equipment can be reduced) and parking management.

This autonomy does not just interest carmakers; information technology companies are also working on specific research and experimentation programs. In this area, Google's ambitions are probably the most 'inclusive', with the vision of a comprehensive service ecosystem oriented towards smart cities, beyond the prototype of the Google Car (and tests carried out on different mass-produced vehicles). Its software program, Flow, is derived from the global management of mobility in a cloud, aggregating multiple data (Google Map and Waze users, wifi terminals, connected parking meters, etc.). The mapping of transport facilities and car parks should make it possible to guide drivers to free spaces (including offering individuals the opportunity to include their private parking space in the system), along with variable parking pricing. Since the software includes all forms of new mobility (self-service bicycles, car sharing, chauffeur-driven vehicles, car pooling, buses and trams, etc.), the objective is to bring cities systems to optimize their transport and usage set-ups. And, once one of the "GAFAs" is able to set itself up as a reference interface between the inhabitants of a city and the authorities organizing public services, most likely putting the latter into competition with partially substitutable services (car pooling, chauffer-driven cars, private car parks), the issue of regulating the power relating to these data will inevitably arise.

6.5 Conclusion: Smart Cities, the Ultimate Stage of Digitalization?

The transformation of the world into an 'archipelago of cities' as the 21st century moves forward, brings as many threats (impact of pollution on public health, financial losses caused by congestion) as it does perspectives for new urban organizations. What the smart cities being formed have in common (in line with the development of telecommunications and the connection of their inhabitants) is the optimization of data management with the aim of improving urban services, and in particular those relating to transport, which includes the threat of 'thrombosis' in major cities. The developments that feed into smart mobilities pertain to urban design and social organization (more compact towns and distance work to reduce flows), and the organization of new ways of managing vehicle and infrastructure capacities (shared fleets, car sharing, urban tolls, road lane management), combined with the mid- or long-term dissemination of incremental innovations (electric vehicles) and disruptive innovations (autonomous vehicles).

For the traditional automotive ecosystem (car and equipment manufacturers), the emergence of smart cities constitutes a potentially disruptive challenge, with the calling into question of internal combustion vehicles in towns, new competition from other industrial players (information technology, services to communities, utilities, etc.) and the diversification of economic models (reliance on big data, less based on ownership, more service-rich).

This phenomenon, beyond the specific issues of transport, also constitutes a powerful factor for changing social organizations and the democracy to run them. Citizens' real-time access to information and their aptitude to co-produce some services (e.g. in a vehicle-to-grid set-up) bring power for action into the public space, with the risk, however, that these new practices might only be taken on by the members of an upper middle class with significant cultural capital. From the moment that these organizations rely on information technology stakeholders (from network operators to social network coordinators), new urban service regulations will need to be invented, in particular since the data environment will create a continuity between public and private spaces, which are also dense with communicating devices.

References

Angelidou, M., (2015), Smart cities: A conjuncture of four forces, Cities, 47, 95–106.

Azari, K.A. et al. (2013), "Evaluation of demand for different trip purposes under various congestion pricing scenarios", Journal of Transport Geography, Vol. 29, Elsevier, Amsterdam, 43–51.

Batty, M. (1990). Intelligent cities: Using information networks to gain competitive advantage. Environment and Planning B: Planning and Design, 17(3), 247–256.

Bélissent, J. (2011). Smart city leaders need better governance tools. Forrester for Vendor Strategy Professionals: Forrester.

Ecube (2010), La « Ville Durable »: Utilities, acteurs des NTIC, grands opérateurs d'infrgastructures : qui tirera les bénéfices de son développement, Septembre.

Hall, P. (2002), Cities of tomorrow: An intellectual history of urban planning and design in the twentieth century (3rd Ed.). Wiley-Blackwell.

Henriot C. (2016), Métropolisation chinoise et villes nouvelles : l'exemple de l'aménagement polycentrique de Shanghai, Géoconfluences, 2016.

International Energy Agency (2016), Energy Technology Perspectives 2016, Towards Sustainable Urban Energy Systems, Paris.

Laterasse, J. (1992). The intelligent city. In F. Rowe & P. Veltz (Eds.), Telecom, companies, territories. Paris: Presses de L'ENPC.

Navarro, C., Roca-Riu, M., Furió, S., Estrada, M., (2016), Designing new models for energy efficiency in urban freight transport for smart cities and its application to the Spanish case, Transportation Research Procedia, 12, 314 – 324.

OECD (2010), Cities and Climate Change, Paris.

Ratti, C., & Townsend, A. (2011). Harnessing residents' electronic devices will yield truly smart cities. Scientific American, 16, August 2011.

Rochet, J.-C. and J. Tirole, "Platform Competition in Two-Sided Markets," 2003, Journal of the European Economic Association, 1(4), 990–1029.

WHO-OECD (2015), Economic cost of the health impact of air pollution in Europe: Clean air, health and wealth. Copenhagen: WHO Regional Office for Europe

World Health Organization (2016), Air pollution levels rising in many of the world's poorest cities, May 2016

Zenetos, T. (1969). Electronic urbanism. Architectural Themes, Greek Architectural Journal, 3, 114–125.

Chapter 7
The Autonomous Car, a Disruptive Business Model?

Danielle Attias

Abstract There is a need to reflect on the autonomous vehicle, a newly emerging type of vehicle, from a historical perspective in terms of the automotive industry as a whole. Rapid and robust technological evolution has made it a tangible concept—taking it from the realm of fiction, into reality. But that's not all. The combined awareness of energy issues and extreme pollution within cities have also participated towards this emergence. Regulations have imposed a de facto restriction of vehicles in cities and the relationship between the user and the vehicle have changed and will continue to change. In major urban centers, the concept of ownership of the vehicle replaces its use. As the relationship to everyday objects turned into customary contract—with the mobile phone, the laptop computer, the idea of becoming a single user of a means of transport is highly seductive; especially in view of the explosion of new transportation choices and the emergence of new business models. Desiring to extend their influence to new clienteles, traditional manufacturers are preparing for the future by investing heavily in the autonomous vehicle. In the US, several states have already authorized the use of vehicles without drivers, as have circuits in Europe. With onboard communications technology, the "smart" car becomes the vehicle of the future, allowing vehicles to communicate and paves the way for a radically new relationship between the user and the mobile object on wheels. Unquestionably, this revolution in mobility poses questions. Future users find themselves ready to adopt this new way of mobility while obstacles both regulatory and technological and social remain. Can manufacturers compete with their new competitors (Google, Tesla, Uber) desperate to meet the challenge of the driverless vehicle and impose new business model?

Keywords Future of the autonomous car · Stakes for manufacturers and users · Technological challenges of autonomous driving · New paradigm of driverless vehicle · Business models of driverless vehicle

---

D. Attias (✉)
Industrial Engineering Research Department/Laboratoire Génie Industriel,
CentraleSupélec, Université Paris-Saclay, Grande Voie des Vignes,
92290 Chatenay-Malabry, France
e-mail: danielle.attias@ecp.fr

D. Attias (ed.), *The Automobile Revolution*, DOI 10.1007/978-3-319-45838-0_7

7.1 Introduction

Kuhn (1962) in *The Structure of Scientific Revolutions* shows how dominant paradigms give way to new ones. For the automotive industry of the 21st Century, this will undoubtedly be the autonomous car. Since 2005, in the Silicon Valley at Stanford University, the first "self-driving car" was born as a result of a project funded by Google. The *Google Car* rapidly became the leader in the autonomous vehicle market, with other players like Tesla, Mercedes, Audi, Apple, and Uber are also determined to break into the market and change the map of the current automotive industry. For Schreurs and Steuwer (2015) "the autonomous driving (self-driving) vehicles, once just a science fiction dream, are a growing reality". If the achievements in this area are very concrete—PSA Peugeot-Citroën announced that it would be able to provide a 90 % of the time autonomous car in 2020—industrial production is still virtually non-existent, with many steps still to be taken. Not only does the self-driving car seem to be the city car of the future—for many reasons which we will present in this chapter, but it is also at the origin of the greatest revolution that the automobile has ever known.

How do we understand the emergence of this new paradigm—the autonomous car? What are the major stakes and obstacles? This revolution of mobility creates a radically new relationship between the driver and his vehicle. How will future users adapt and appropriate this *moving object on four wheels*? What are the resistances and difficulties to anticipate so that the autonomous car quickly and comfortably takes its place in the connected city area?

The current vehicle automation, which is the beginning of automated driving, includes many challenges and progressively leads us to the era of the self-driving car, in a world where everything automated has become commonplace (Sect. 7.2). Many obstacles remain however to be overcome—such as social, legal and financial issues—to be taken into account by manufacturers in this market (Sect. 7.3). Finally, a new area of mobility forms between the *object*—the autonomous car—, the *users*—interconnected drivers—and the *ecosystem* within urban centers will promote these interconnections (Sect. 7.4).

7.2 The Automation of Vehicles: An Underlying Trend

7.2.1 Embedded Technologies, Beginnings of Automated Driving

The Google project undeniably accelerated the entire automotive market. We must not however, forget all the technological advances experienced by cars in recent years. The connected car would be the missing link between the completely mechanical car of yesterday, and the fully autonomous car of tomorrow.

The mechanical car of the past might now more easily find its place in collectors' garages. With the 1973 crisis and the shortage of oil, the automotive industry has had to profoundly revisit the issue of car production—which must be redesigned to consume less and last longer. In Europe, diesel is becoming commonplace, along with embedded electronics. Injection is controlled electronically, but not only. Gearboxes managed by computers are also appearing, active safety with the appearance of driving aids such as ABS, which became standard in the 80s, or ESP. These are the beginnings of automation that transform the user experience, while at the same time change the car from a transportation tool to a place of comfort and convenience.

But these few past developments were only the beginnings of the connected car of today. The technologies used now are numerous and affect all areas of driving, in order to constantly improve the pleasure for the passenger. Valeo, for example, has developed a system allowing the car to park all by itself, where the user can leave his vehicle to park and come back by itself once needed again. The advantage of this technology is also gaining an important place in car parks. In the absence of this technology, the space left between each vehicle is calculated with a side door opening to the right and to the left.

This empty space needed for opening doors makes parking spaces unavailable to an average of one-third of vehicles which, with this technology could parallel park, side by side, without any loss of space. This system was developed in Japan, in the Tsukuba Robotics Lab, where the question of the lesser m^2 gain is crucial. But it does not stop here. Connectivity also takes place through the navigation systems—GPS, maps, traffic information, etc.—, The media—radio, MP3 music, photos and videos—, the phoning—Bluetooth connection, call management—driving aids—rearview camera, onboard computer, fuel consumption—, or even services—support, weather … Ultimately, it is the driving itself that will be the subject of this hyper-connectivity and will profoundly change our traffic patterns.

The new SAE International Standard published in 2014 is a report with a taxonomy for vehicle automation, created in order to provide a common language to all manufacturers for automated driving and to establish a harmonized classification system (SAE 2014). Six driving levels are identified from the level of lack of automation to full automation. Indeed, today it is necessary to identify and define the level of automation of road vehicles because users have to perform specified driving tasks depending on the vehicle automation level.

If the automation of cars follows the progress of information systems, and the increasing integration of electronics in vehicles is not yet complete, it is thus fitting to describe the corresponding stakes.

7.2.2 *Many Stakes for Manufacturers and Users*

There is no doubt that the automated car, which remains today at the level of the connected car, represents many stakes for carmakers and for society as a whole.

The first issue, a major one for manufacturers, is the market issue that justifies, probably alone, their involvement in such a niche. According to a study by Bosch, "nearly a driver out of three is willing to buy an automated vehicle, able to steer, brake and accelerate on its own initiative". There is a real trend for connected technologies and automation is only an extension of this phenomenon. Carmakers must thus take into account their consumers evolution in order to offer adapted cars for their needs—that is to say, with modern technological content in line with the technologies that exist and used by them at home. But it is not just about offering cutting-edge technologies for consumers, it is also proposing a transport mode that is coherent with population evolution. Thus, according to the US Census Bureau, the over-65-year demographic/category will represent approximately 28 % of the EU population and 20 % of the North American population by 2050. This is a very big challenge for carmakers because these older people have for some, mobility difficulties, coordination or simply perception and automated vehicle is a way to overcome these many daily obstacles.

The corollary of this aging problem is the issue of security. With an increasing number of older people on the roads, under certain conditions, road safety will be more difficult to ensure. The progressive automation of vehicles opposes then a technological response to biological human determinism. The vehicle allows these people to drive—or not to drive and therefore be transported—in peace and security. This car will adapt its speed to others through information exchange in real time with vehicles located around and takes care to respect the speed limits at any time, any place—whatever the weather conditions. But security is not the sole concern of the elderly. The driving car of tomorrow will have a consciousness of its environment that is much more advanced than humans due to its connectivity, and will be at the service of all types of users. The SVA project (Simulation for Autonomous Vehicle Safety) developed by the IRT SYTEMX (Technological Research Institute of Paris-Saclay) has set the goal of answering this question using numerical simulation, by developing methods and support tools for design and validation, for the challenging complexity of both the large number of situations that the driver meets on road and technologies for vehicle automation.

The last major issue, from the perspective of the cities this time, is traffic management. The massive influx of cars to urban centers in the morning or towards the periphery in the evening regularly cause traffic jams. The automated car is a way that can be simply used to improve traffic flow. In case of slow traffic, cars that are located towards the front of the road will receive information about traffic problems and will reduce their speed and will have immediate effects not to cause jams, allowing better traffic flow even when slowed. Studies are developing in this direction such as the project conducted in 2015 by the IFSTTAR (French Institute of Science and Technology, Transport, Spatial and Networks) (Billot 2015), which developed TEMPUS, a software platform for computing multimodal itineraries including all the components such as the dynamic courses times and the interface times (this notion echoing the one of modal shift). Obviously the self-driving car will make fluid all this traffic and prevent many accidents, because of its highly secure system. The study by IHS Automotive, *Light Vehicle Production Forecasts*

(IHS Automotive 2015), goes in this direction and considers that "*the number of accidents will be close to zero for these autonomous cars, road traffic could be regulated*". The highways adapted to these vehicles of the future are already being designed. The SARTRE project—Road Trains for the Environment—(2016), aims to develop technologies allowing vehicles to drive and change and their impact on the environment. Within the Energy Technologies Area (ETA) of the University of Berkeley, Greenblatt and Shaheen (2015) study the relationship between the demand for mobility and the impact in terms of cost, public health and the environment. The project done by Greenblatt and Saxena (2015) shows that while improving security and convenience, autonomous taxis can significantly reduce greenhouse gas emissions. They published a prominent paper in Nature Climate Change in 2015 on the greenhouse gas benefits of autonomous taxis. They continue to explore other innovative technology ideas, like opportunities to improve the net energy performance of photo electrochemical water-splitting technology.

The future of car automation can no doubt be seen in connected cars whose embedded technology is increasingly sophisticated and requires less and less human intervention. This automation is even more important, that it integrates consumer issues, security and pleasure in using this transport. There is no doubt that all these recent developments gradually lead us to the era of the self-driving car, whose technology seems already to be within reach.

7.3 The Beginning of the Era of Self-Driving Vehicles

7.3.1 Developing Technologies

As we have seen, automation is a trend in the automotive industry since its beginnings. This trend seemed to run up against the automation of the last essential controls that are accelerating, braking and steering, which today remain the sole responsibility of the driver. Nevertheless, we note that some recent marketed innovations begin automating these last three controls.

For example, Adaptive Cruise Control has developed a system, now commercialized by Audi, which automatically regulates the speed of the on-road vehicle, depending on the distance between the vehicle and the one ahead. This is a first automation for speed control, more sophisticated than the cruise regulator. In the case of braking controls, the Collision Warning with brake support by Ford is one of the first systems to automate the braking of the vehicle. The system detects if the vehicle has to stop suddenly, warns the driver and failing a response from the driver, the system itself takes care of braking. Finally, automatic parking is the first fully automated system marketed by many manufacturers. The system allows parking without the driver touching any of the vehicle's controls. Among the manufacturers who market this technology are Toyota, Ford, BMW, Mercedes and Audi.

Thus, the latest features to be automated which are the basic controls of the vehicles—the direction, acceleration and braking- are already part of a mainstream automation, as attested by the mentioned technologies above. But the step from partial automation and full automation remains important.

What about then the total cars automation? For IEEE (Institute of Electrical and Electronics Engineers), “75 % of cars circulating in 2040 on the US territory are autonomous vehicles”. The IEEE goes even further, the institute predicts how infrastructure, society and even behaviors may change when self-driving cars become the norm on our roads (Solyom and Coelingh 2012).

So far Google is the first actor to have started seriously in the race to automation. The system developed on the Google Car is based on five broad sets of sensors: a camera, radars, a GPS receiver, sensors on the wheels and, finally, a remote laser sensing system. This system is intended to enable the autonomous operation of the car—automatic driving—by processing data signals received on the road without removing control from the driver, who may regain it at any time. Until now very theoretical, the Google Car has succeeded brilliantly the testing phase, by installing the automated control system on eight vehicles which have completed more than 200,000 km in California without encountering the slightest incident.

This development project was launched in 2005 is currently being tested in several US states. Since 2012, over one million five hundred thousand kilometers were covered and only one minor accident was noted, caused by human error. The Google Car is ready and approaches 100 % automation. However, it is still limited by exceptional conditions such as extreme weather or the particular case of managing firefighting sirens. It seems indeed difficult to make the Google Car intelligent and able to handle any situation, that is to say any hazard. However, for nearly 95 % of the situations encountered by users, this is possible since the Google Car has reached this level of automation.

The competition was quick to follow the direction given by Google. Two types of actors emerge on this new technology, actors in the automotive industry such as Mercedes, Tesla, and emerging players, out of the automotive industry like Apple and Uber. Mercedes is well advanced since the group presented a prototype for the CES (Consumer electronics show) in Las Vegas in January 2015. This model, the F015, is positioned as both independent and visionary, with an interior fitted out as a living room. It is actually the car of tomorrow. Tesla (Génération-NT 2015) and other carmakers such as BMW, Ford etc. also have ambitions of mastering the technology to make it accessible to the general public. However, they are lagging behind Google and Mercedes as they have no prototype yet. Similarly, Uber justified its interest in the development of this technology but so far less advanced compared to the leaders, Mercedes and Google.

7.3.2 New Business Models

If the presence of traditional actors of the automotive industry such as Mercedes, BMW, Ford in this segment is not surprising, given the high level of competition and innovation that characterize this industry; then the presence of new players like

Google or Uber is all the more surprising. The strategic reasons for these players to invest in the autonomous car are still not revealed, but they present heavy implications for the future of the automotive industry.

A study conducted by KPMG in partnership with the Center for Automotive Research (2014a) shows that **a business model must be reinvented for carmakers**: the advent of the autonomous car allows us to imagine vehicle fleets for collective use where the user would pay per kilometer instead of buying his car. This implies a significant change in the way cars are financed today, especially as the automotive park should logically be reduced significantly. These fundamental changes will be further amplified by the increased importance of information systems in the automotive value chain.

Uber sees in the self-driving car a potential disruption of its business model and market. Indeed through the use of self-driving taxis (the fee paid to drivers being a major cost item for Uber) it could dramatically reduce costs and consequently prices. Thus the Uber service could very well be competitive compared to the total cost of car use by the consumer. In other words, the way to consume cars would be in complete disruption with what we know today and Uber would be at the heart of this system.

Regarding Google, the group has the strike force and the DNA to engage in such an innovative project (KPMG 2014b). First, with its mapping technology—essential for driving automation- Google is one of the most relevant actors to develop this technology. Second, through the automation of vehicles, the group can see a potential proliferation of data movement from its users, data it knows how to valorize well. Finally, having control over the vehicle itinerary, Google would be in a strong position to suggest nearby services (Alternatives Economiques 2015).

Thus, the era of self-driving vehicles is undeniably upon us. The technologies are already developed, and many actors are interested in the subject, whether they are historical actors of the automotive industry, or new players ready to shape their way into the automotive industry of tomorrow. The industry is changing rapidly, but there are still a number of barriers to overcome such as the legal barriers, or the marketing of such a disruption.

7.3.3 Existing Barriers

7.3.3.1 A Legal Framework to Adapt

With the automatic parking, ABS or automatic lightning, automation is already present in modern cars but in a light and targeted way. In all cases, the driver must keep control over his vehicle at any time as required for the moment the European legislation. The autonomous vehicle is not legal on public roads but nonetheless permitted in private areas. In France, the Vienna Convention of 8 November 1968 states that "every moving vehicle [...] shall have a driver", "the driver must always be in control of his vehicle" and "the driver of a vehicle must avoid any activity other than driving". These principles aimed at minimizing the risk of accidents therefore oppose the full automation of driving. The signatory states including France began discussions for a less strict application of this article. The United States, which did not sign the Vienna Convention is therefore not restricted. Thus, some states already allow the movement of such vehicles without drivers. The DMV (Department of Motor Vehicles) drafted this law which lays down two conditions:

- Autonomous vehicle manufacturers must approve the car by the DVM
- Drivers must obtain a different license and are obliged to stay behind the wheel in case the service encounter a problem.

In Europe, the change is accelerating with the decision of the UK Department of Transport and the Minister of Trade and Innovation to launch autonomous cars to be tested on public roads. An amendment of the Vienna Convention is scheduled for 2017 and will allow these vehicles to operate in France. They can be driven alone but, like in the US, the driver must be behind the wheel to manage "exceptional" situations such as the roads under construction that these autopilot systems cannot yet detect.

In France, the research institute VeDeCoM (Véhicule Décarboné Communicant)[1] is developing technological bricks meeting the driving delegation issues in urban and peri-urban autonomous vehicles. The institute develops safe functional architectures to drive an autonomous vehicle with high levels of reliability, artificial intelligence algorithms to enable the autonomous vehicle to make decisions, means of testing to simulate and test demonstrators in critical situations and validate the associated solutions. Finally, the Institute aims to integrate multiple technologies from carmakers, equipment suppliers, but also research laboratories (VEDECOM 2015).

7.3.3.2 A New Concept of Responsibility

Similarly, this new way of moving generates some questions regarding the responsibility, especially in case of accident. For (Schreurs and Steuwer 2015) "*the Zero-Accident-Vision has been important message for developers of autonomous vehicles and component suppliers. The vision appears to play a larger role in the United States where there are higher fatality rates than is the case in Europe or Japan*".

---

[1]**The VEDECOM Institut** (Véhicule Décarboné Communicant) is one of the IET (Institute for Energy Transition) established under the Investment Plan for the Future of the French government. Its research involves multidisciplinary work involving physicists and chemists, mechanics and electrical engineers, electronics engineers and computer scientists but also sociologists, psychologists, economists and legal experts to study the impact and acceptability of new use cases and the new ergonomic and regulatory arrangements to put in place.

In France, the law of "Badinter" of 5 July 1985 states that the passengers of the vehicle, the driver excluded, should be compensated for the damage suffered in the event of an accident. It could be said that the majority of the damages implicate a single responsibility from the driver. In a criminal level, the Highway Code also states that it is the driver who is responsible. One may then wonder who is responsible in case of use of a driverless and fully autonomous vehicle. Insurance, which compensate in case of road accident, will then return to the carmakers in case of failure in the automatic control of the vehicle or to the IT services companies, or even towards the responsible authorities on road maintenance. What is certain in this widening circle of responsibilities is the legal conflicts that invariably will be born out of this complexity.

For lawyers, the responsibility will be initiated if and only if it is proved that the damage is due to a fault in automation. The criminal responsibility of the driver is much more difficult to define. Indeed, if the driver has the opportunity to pursue other occupations while the car is moving, we cannot impose on him the obligation to correct any autopilot flaws. If, however, the driver is obliged to intervene in case of a technical problem, criminal liability may be incurred in an accident and even if the accident was caused by a system failure.

However, some insurance companies see in the arrival of the autonomous car, a positive legal assessment element. In fact, in the video that would be produced—since in the autonomous car, everything can be filmed- the ability to better analyze the causes of the accident from the current situation is real.

7.4 The Human at the Center of Concern

One of the barriers to the use of autonomous vehicles is still the price. For much of the population, any communicating vehicle and containing many technologies will have a very high purchase cost. Indeed, in 2012, Google announced that its vehicles would contain over 150,000 € of technology. This barrier remains only psychological because automakers and other players have understood that, and are doing everything possible to reduce this price and thus change that image with their future customers. A vehicle, autonomous or not, remains a vehicle that must remain accessible if one wishes it to become widespread.

7.4.1 The Autonomous Car and Acceptability by the User

A paradigm shift from the traditional car to the autonomous car is emerging, despite ownership models still being linked to social status and the desire to possess. This is certainly the case in certain emerging countries (China, India and Latin America) and in certain social categories of industrialized and emerging countries who see the traditional car, a social success of representation, individual liberty and power.

However, another option has been emerging for some time regarding the use of the vehicle in which the idea of property is changing and has led many users in large cities preferring cars to rent than to buy. Many factors militate in favor of this choice: firstly, the abundant supply of new services offered by manufacturers and car sharing companies that facilitate these new travel choices. Added to this is an evolving digital environment simplifying the management of travel choices? People being able to manage the different movements of their day from home via a mobile device has become commonplace and comfortable. In addition, cost reduction is an important factor because only the time of use is chargeable. Finally, convenience should be stressed as we get a vehicle in good condition without having to manage the maintenance and management of its transport mode is optimal since it is possible to rent a two-wheeled vehicle on weekdays and choosing a sedan-type vehicle for family trips on weekends.

In future scenarios that we can look forward to the autonomous vehicle as a car-pooling solution. If the rental car is becoming commonplace these days, then it becomes possible to think of the autonomous car in terms of shared use as well. This means that this vehicle would first be a car that meets the mobility needs of people and services could be shared by residents of a neighborhood or a city. Nobody is owner of this car, it is managed by an autonomous vehicle leasing company that would take over the complete management of this type of vehicle (insurance, maintenance, remote management of time and use). Only the time used would be paid for by the user. But sharing the same vehicle at certain times of the day is also an opportunity to create social bonds and live in interaction with others. Perhaps, new communities would emerge among users of cars, sharing companies or carpooling.

In this type of scenario, the autonomous car becomes an intelligent product and a proactive service on four wheels. Thus, the use of the car requires the services of a robot. We could ask the car-robot to pick us up at home, walk the dog to the park, to wait while it plays and bring it home before moving on to take pre-ordered shopping at the supermarket and finally accompany us to our appointments.

However, in order to modulate the adoption curve of this new tool of mobility, it is also necessary to question the acceptability of future users. Are we ready for such an alternative?

7.4.2 From Ownership to Vehicle Use

Another option has been emerging for a while regarding the vehicle use in which the idea of property is changing and leading many users to prefer renting a car rather than buy it. Many factors contribute to this choice: cost reduction since only the time of use is priced, convenience because the vehicle is obtained in good condition without having to manage maintenance, simplified management of its transport mode since it is possible to rent a two-wheeled vehicle on weekdays and choose a sedan type vehicle for family trips on weekends.

To modulate the adoption curve of this new mobility tool, it is also necessary to take into account the simple use of the vehicle and the human relationship to the object. The results of an international study conducted by Kyriakidis et al. (2015) on automated driving has helped to clarify the expectations, the acceptance and the fears of users on this type of vehicles. 5000 people from 109 countries around the world have responded to this questionnaire. The fully automated driving was rated "very good" for 33 % of the interviewees. 69 % of respondents said that the fully automated driving will reach a global market share of around 50 % before 2050. Regarding the areas of concern for the future users, the main points raised in this study are the issues related to the embedded software piracy, legal issues and misuse of transmitted data. Another global survey conducted in 2015 by BCG (Boston Consulting Group) and the World Economic Forum from 5500 consumers in 10 countries (Asia, North America, Europe, Middle East and Asia) highlighted the fact that the users who adhere most to the autonomous car are more numerous in emerging countries such as India (85 %) and the UAE (70 %) than in France (60 %) and Germany (44 %) or Japan (36 %). Another point to note in this survey is that, traditional manufacturers are by far those that consumers consider most reliable (between 50 and 58 % confidence) and think about other technology players like Apple and Google as bringing "relevant expertise…but complementary" (BCG and WEF 2015).

A critical mass of the population will be constrained in the purchase act because they don't experience the "driving pleasure". Manufacturers and other actors in this development must then have a much broader vision and thus "compensate" the apparent lack of action from the vehicle occupants. The idea is to make that time more productive than it is today. For this, they need to strategize on the equipment that may be contained in the automobile. The main idea is to be able to work, interact with other during this time that was previously considered as wasted time. A second barrier which could be difficult to overcome is the one of physical security. The fear of the new and the irrational fear of not being in control still remain present for a portion of the population. Indeed, there are behaviors of resistance, skepticism from certain categories of people towards this technology. This certainly reveals a mistrust of technology in general, but in the case of the autonomous car this wait and see attitude requires that manufacturers have solid arguments, concrete evidence for secure driving in any place.

7.4.3 The Autonomous Car in a Connected Ecosystem

Automated vehicles require active planning from local authorities. Indeed, the local public actor must consider the impact of automated vehicles on urban planning, urban transport and the design of future infrastructure projects. This is creating a new and true mobility ecosystem. According to Jean-Louis Marchand, Deputy Director of Eurovia (2014) "The connected vehicle will have to find its use in relation to other infrastructure uses". In addition, the road networks are not managed in a unique way: 20,000 km of national roads are managed by the Ministry of Ecology, 380,000 km of provincial roads are managed by departments and 650,000 km of local roads are managed by municipalities. There is not a real coordination structure between these different managements. "*We should be able to say in the years to come in which network it may be accepted of a particular type of vehicle with a particular level of intelligent service and we will not escape this reflection*" says Jean-Louis Marchand. To completely replace the driver with a communicating robot, one should also develop the "smart" road, with powerful communication infrastructures and networks to connect the different vehicles.

This can also have an impact on many other sectors of the economy such as the management of public car parks or simply the public road security officers. This change will also impact the manufacturers of the automotive sector itself because all the actors must work together to implement a solution with a "common language". For this, they gather in consortiums such as "Car 2 Car communicating" with the aim of encouraging the creation of European standards for "intelligent" vehicles that would be those of all manufacturers in Europe. Two new forms of communication are emerging: the communication from vehicle to vehicle (V2V), and the communication vehicle to infrastructure (V2I). V2V communication allows cars to share their real-time positioning in order to avoid collisions or optimize the driving of a—restricted—group of vehicles. V2I communication allows vehicles this time to share their position, but also the destination and the route they want to take with a central unit, whose role would be to coordinate and dispatch the traffic information and itineraries.

The explanations of Alberto Broggi, member of the IEEE (Institute of Electrical and Electronics Engineers) and Professor of Computer Engineering at the University of Parma allow once again to clarify these concepts: "*Imagine that all cars are connected and that a central unit can know precisely their position and destination. The central unit can send speed adjusting instructions for the vehicles approaching an intersection, so they do not collide and they cross the intersection each on their turn, optimizing their movements. In this case, the traffic lights will not be needed, coordination being managed at a higher level*".

Tests are already underway in Europe (especially in Germany) and the United States, combining the two communication systems V2V and V2I, a global type of communication called V2X. The IEEE forecast includes the future disappearance of signs and traffic lights, unnecessary for autonomous cars since they will possess or they will be communicated in real time all the necessary information (speed limits, traffic direction). Moonzur (2012) also mentions the likely disappearance of the driving license. "In case the vehicle would be completely autonomous and no driver intervention is allowed, it is the vehicle that will be in sole command". Indeed, no need for qualifications or special knowledge, and so no need for a driving license that would become completely useless. But another consequence is noteworthy, autonomous vehicles should allow access to disabled persons with more or less serious handicap.

These economic issues can be solved if all the stakeholders at national, European and global levels manage to cooperate together. The main barrier becomes therefore time because all these measures cannot be quickly and easily implemented.

7.4.4 The Autonomous Car in a Period of Transition

One question cannot be overlooked for autonomous vehicles and their release is the more or less long period (from manufacturers from 15 to 20 years) where the latter will coexist with conventional vehicles. It is clear that different levels of connectivity will involve different driving modes. Speed has already been highlighted as a potential obstacle to cohabitation. Indeed, the autonomous car is connected to software which is integrated in an absolute respect for the Highway Code. Its slow speed is presented as counteractive to performance, especially in a "crowded" environment in vehicles. The autonomous car, when too slow, can become a danger to other cars, especially as their numbers can increase rapidly in urban centers. Another question is that dedicated lanes for these cars that complicate traffic, traffic management and security because it must be noted that the major accident of Google (March 2016) was in connection with a collision due to an overflow of reserved lane.

But in July 2016, "after the death of an autonomous car driver in the US, the reputation of these vehicles considered the future of the automobile has taken a hit". It is in these terms that many international newspapers (Courrier International, Techcrunch) have reported the first fatal accident of a driver of a Tesla Model S, equipped with an Autopilot system that "neither the system nor the driver detected the operation of a truck and were therefore unable to activate the braking system". This accident has weakened the defenders of this technology and points to an ethical debate on the choices to be made, especially in terms of the inevitable danger with a pedestrian, for example.

Researchers from the University of Toulouse, Oregon and MIT published the results of a study on the choice of autonomous cars facing danger of the road. These cars "sometimes have to choose between two evils, like flipping pedestrians or sacrifice themselves and their passengers to save these [same] pedestrians". They asked 2000 Americans to choose which among pedestrians and passengers have to be sacrificed if an autonomous car can avoid a catastrophic collision. The result is largely in favor of the pedestrian, "75 % of respondents think so that the autonomous vehicle should, in extreme cases, to turn from the pedestrian, even end in a wall and kill the driver…". This new episode in construction raises the question—is the autonomous car to slow down the development of this technology? Nothing is certain, because as written by Mary Cummings, head of the laboratory of autonomy at Duke University "my fear is that this accident could have been avoided" (AFP 07/02/2016) and regrets that the Autopilot system of Tesla not been sufficiently tested.

This question highlights the cognitive limitations of our future cars-robots that can adapt in real time to unforeseen or unexpected factors (e.g. recognition of flying objects or identifying a new problem). Another major risk, these cars can misinterpret the behavior of pedestrians. In the case of children (who often stop in front of a ball that rolls on to the road and then run to catch it), poor anticipation can be fatal. The unpredictability of our behavior is not yet sufficiently taken into account by the existing software. The next step will be a more reliable adaptation of technology of the car to the environment and the convergence of two driving modes.

7.5 Conclusion

The automotive industry has always evolved with numerous innovations whose purpose is to improve user experience, security and comfort. Current advances in robotics and in digital technologies enable the automotive industry to achieve vehicle automation. Indeed, the development of numerous sensors and artificial intelligence allow today for the automation of certified and tested cars on the roads. The first prototypes such as those from Google and Mercedes have shown to be convincing in the face of real-world conditions.

The stakes behind this disruptive innovation are numerous. First, the strong competition in the automotive industry always pushes manufacturers to innovate and propose new mobility and "driving" experience adapted to 21st century consumers. Autonomous driving provides solutions to a number of issues posed to manufacturers: comfort, safety or traffic. Next, another major stake concerns the context of our society digitalization that is gradually transforming many industries. Automated cars certainly translate as the digitalization of the automotive industry, including the arrival of new players like Google and Uber, players who are ready to create disruptive business models. Moreover, in a medium term perspective, this evolution will not stop at the autonomous car and opens the field for the design of similar technical objects. One can imagine the autonomous bus or trucks from the same model as the car, and at a later stage, autonomous boats and aircrafts.

If many obstacles, both legal and human remain, the dream of using a self-driving car today is closer to reality and opens the way to a societal reflection on the future of mobility.

Acknowledgments This research benefits from the support of the Chair "PSA Peugeot Citroen Automobile: Hybrid technologies and Economy of Electro mobility", The Armand Peugeot Chair led by CentraleSupélec, ESSEC and sponsored by PEUGEOT CITROEN Automobile.

Bibliography

Alternatives Economiques, 2015. Google veut révolutionner la mobilité [WWW Document]. URL http://www.alternatives-economiques.fr/google-veut-revolutionner-la-mobilite_fr_art_1345_71338.html (accessed 3.19.16).

BCG, WEF, 2015. Revolution in the Driver's Seat: The Road to Autonomous Vehicles.

Billot, R., 2015. On open-source multimodal trip planner, IFSTTAR. Presented at the ITS congres.

Eurovia, 2014. Rapport d'activité 2014.

Génération-NT, 2015. Elon Musk : une voiture Tesla complètement autonome, c'est dans deux ans [WWW Document]. Génération-NT. URL http://www.generation-nt.com/tesla-motors-elon-musk-voiture-autonome-calendrier-actualite-1923122.html (accessed 3.19.16).

Greenblatt, J.B., Saxena, S., 2015. Autonomous taxis could greatly reduce greenhouse-gas emissions of US light-duty vehicles. Nature Clim. Change 5, 860–863. doi:10.1038/nclimate2685

Greenblatt, J.B., Shaheen, S., 2015. Automated Vehicles, On-Demand Mobility, and Environmental Impacts. Curr Sustainable Renewable Energy Rep 2, 74–81. doi:10.1007/s40518-015-0038-5

IHS Automotive, 2015. Light Vehicle Production Forecast. USA.

KPMG, 2014a. Global automotive executive Survey with the Center for Automotive Research.

KPMG, 2014b. Self-driving cars : The next revolution.

Kuhn, T.S., 1962. The structure of scientific revolutions. University of Chicago press.

Kyriakidis, M., Happee, R., de Winter, J.C.F., 2015. Public opinion on automated driving: Results of an international questionnaire among 5000 respondents. Transportation Research Part F: Traffic Psychology and Behaviour 32, 127–140.

Moonzur, R., 2012. Les voitures autonomes, bientôt sur nos routes ? Techniques de l'Ingénieur.

SAE, 2014. Taxonomy and Definitions for Terms Related to on-road Motor Vehicle Automated Driving Systems, J3016.

Schreurs, M.A., Steuwer, S.D., 2015. Autonomous Driving – Political, Legal, Social, and Sustainability Dimensions, in: Maurer, M., Gerdes, J.C., Lenz, B., Winner, H. (Eds.), Autonomes Fahren. Springer Berlin Heidelberg, pp. 151–173.

Solyom, S., Coelingh, E., 2012. Performance limitations in vehicle platoon control. Presented at the 2012 15th International IEEE Conference on Intelligent Transportation Systems, ITSC 2012, Anchorage, AK, 16-19 September 2012, pp. 1–6.

VEDECOM, 2015. Rapport de l'Institut VEDECOM (Véhicule Décarboné Communicant et sa Mobilité), ITE (Institut pour la Transition Energétique) mis en place dans le cadre du Plan d'Investissement d'Avenir du Gouvernement Français.

Part III
Technical Challenges to Overcome Towards Electromobility

Chapter 8
Is Electric Vehicles Battery Recovery a Source of Cost or Profit?

Hakim Idjis and Pascal da Costa

Abstract The lithium-ion battery technology as the today's best available technology is a key in accompanying vehicles electrification. Its end-of-life recovery is lever in overcoming technical challenges towards electromobility deployment, such as battery cost, environmental impact, the availability of constituent materials and the mandatory recycling rate. In this chapter, we focus on economic aspects, in order to assess the end-of-life recovery impact: we analyze the end-of-life cost evolution of lithium-ion batteries to determine whether it will be a source of cost or profit for car manufacturers. We define and analyze two recovery options: on the one hand, simply recycling which is mandatory by regulation and, on the other hand, repurposing for reuse in many second life applications (from residence related applications to energy storage and grid stabilization). To account for the complexity and the long-term horizon of our study (2030), we combine the use of System Dynamics with the Stanford Research Institute Matrix for building scenarios that mix relevant factors such as the electric vehicle market and the proportion of repurposing for reuse. Finally we show that repurposing could lower the battery's initial cost—under certain conditions regarding the future battery price and the repurposing cost—where recycling might increase it.

Keywords Electric vehicle · Lithium-ion battery · Recycling · Repurposing · 2nd life

8.1 Introduction

The complexity of the transition to electromobility raises questions about the contributing factors, such as energy, socio-economic, political and technological ones. The latter is the topic of this third part of the book. In this chapter we tackle

H. Idjis (✉) · P. da Costa
Laboratoire Genie Industriel, CentraleSupélec, Université Paris-Saclay,
Grande Voie des Vignes, 92290 Chatenay-Malabry, France
e-mail: hakim.idjis@ecp.fr

P. da Costa
e-mail: pascal.da-costa@ecp.fr

D. Attias (ed.), *The Automobile Revolution*, DOI 10.1007/978-3-319-45838-0_8

the battery end-of-life whereas in the next one we will investigate firstly the possibilities offered by electric fleets to provide valuable services to the electrical grids, and secondly how the existing rules need to be redesigned in order to expand this type of new business models.

In the context of climatic change, many countries implied in the Paris COP21 in December of 2015 have announced national targets in reducing the emissions of greenhouse gases. These targets will be revised upward every five years according to various criteria, including the evolution of the technical change. Various technological roadmaps (ERI 2009; IEA 2012, 2014; IPCC 2014) from which the governments try to asses some choices and trade-offs consider as crucial the electrification of the transportation sector via the deployment of electric vehicles. Indeed the transportation sector, road transport in particular, causes of many negative externalities today due to the dependence on oil (European Commission 2014a). In France, over one third of CO_2 emissions (the powerful greenhouse gas) was due to transportation in 2010, with road transport accounting for 96 % of the transport emissions. In the United States of America, about 30 % of CO_2 emissions and a share of 86.4 % for roads in 2010. In China transport accounted for only 7 % of total emissions in 2010, but with a fast growth rate of the number of vehicles of exactly 11 % in 2010, transport and especially road transport will be increasingly huge for future Chinese CO_2 emissions. Therefore it proves critical for every country, whatever its level of economic development is, to develop electric vehicles together with the decarbonization of the mix of electricity in order to mitigate global warming (Alazard-Toux et al. 2014; ERI 2009; Saveyn et al. 2012; Tian and Da Costa 2014).

In the climate package of the European Union (EU), commitments call for a transport sector reduction of 60 % in 2050 compared to CO_2 emission levels in 1990 (European Commission 2011) and emission thresholds for new light vehicles (personal cars) are defined up to 130 g of CO_2 per km in 2015 and 95 g per km in 2020 (European Regulation No. 443/2009: European Commission 2009; European Regulation No 333/2014: European Commission 2014b). While short-term goals seem attainable by the various car manufacturers (EEA 2015) (the goal of 2015 was completed), those of long term are real challenges and will require radical innovations thanks to the new Electric Vehicles (EV). Indeed fuel cells and other promising technologies being less mature in terms of technology and infrastructure, the industry tends to turn towards EV: hybrid, plug-in hybrid and battery-electric (McKinsey and Company 2010). These EVs mainly use Lithium-Ion Batteries (LIB) that give them greater autonomy. However they crystallize some of the barriers preventing widespread use, such as the cost of the battery, its impact on the overall life cycle assessment of the EV and the availability of constituent materials. Moreover, EU regulation imposed on car manufacturers the recycling of at least 50 % of the battery weight (European Commission 2006). The investigation of the battery end-of-life appears then necessary for the following reasons: lowering the battery/vehicle cost and environmental impact, control of the supply of potentially critical materials and compliance with the regulatory recycling rates.

In this chapter we will analyze the end-of-life cost evolution of LIB to determine whether it will be a cost or a profit for car manufacturers. These latter already raise this question in order to guide their strategies (the research chair on electromobility "Armand Peugeot"/CentraleSupélec and ESSEC, France, which supported our research work is a good example). To investigate our research question, we will define and analyze two recovery options: (i) The recycling which is mandatory by regulation; (ii) The repurposing for reuse in second life applications (from residence related applications to energy storage and grid stabilization: see Törkler 2014). The end-of-life cost is the sum of the costs of these two options: recycling being mandatory and repurposing will take place before recycling if profitable. To account for the complexity and the long-term horizon of our study (i.e. 2030), we combine the use of System Dynamics (Sterman 2000) with the Stanford Research Institute Matrix for building scenarios. Our set of six techno-economic scenarios show that it is possible the recovery could not lower the initial LIB cost, or even increase it by 26 % in the worst case; And in the case when it is possible to reduce this initial cost, it is the repurposing that would contribute to it, even if this initial cost reduction is weak (11 % of the initial cost in the best case).

Section 8.2 now introduces the technical scope about battery and recovery technologies: these concepts are necessary for understanding the subsequent analysis. Our approach and results will be detailed in Sects. 8.3 and 8.4 respectively. Section 8.5 concludes this chapter.

8.2 Battery and Recovery Technologies

8.2.1 Battery

For vehicle electrification needs, several battery technologies are available: nickel cadmium (Ni–Cd), nickel metal hydride (Ni–MH) and lithium-ion (LIB). Nevertheless LIBs are more adapted and used (energy density, voltage of cells, lifespan and memory effect). A LIB is made up of the electrochemical (set of modules, which are composed primarily of cells) and the electrical and support part (battery management system, connecting cables and cooling system) (see Fig. 8.1).

Therefore the basic component in a battery is the cell. A cell consists of an electrolyte, a separator, an anode (negative electrode) and a cathode (positive electrode). Each of the two electrodes is composed of a conductor and an active material. The negative active material of the anode is usually graphite. The positive active material of the cathode is a combination of **lithium and a metal oxide**, which varies from one technology to another. It is this latter that defines the properties of the LIB (lifespan, safety, capacity, and cost) (and gives it its name). In this chapter, we selected two families of key technologies:

- Nickel-Manganese-Cobalt based batteries (NMC) with the average composition $\mathbf{Li(Ni_{1/3}Mn_{1/3}Co_{1/3})O_2}$ (ADEME 2013; Hoyer et al. 2014);
- Iron phosphate based batteries (LFP) with the composition $\mathbf{LiFePO_4}$.

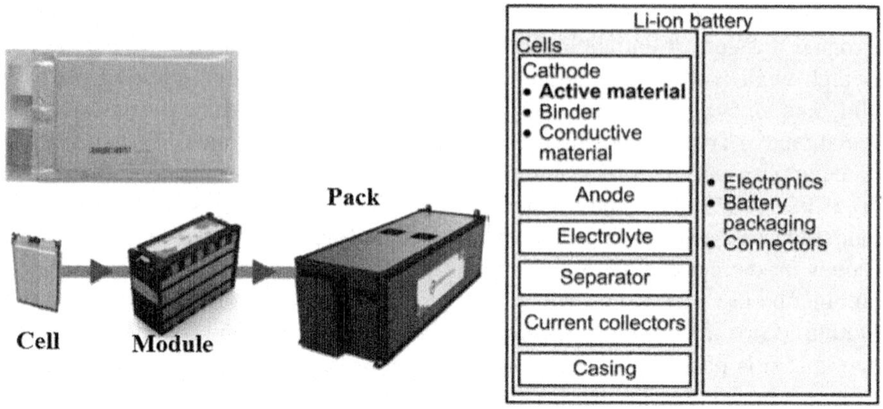

Fig. 8.1 Composition and decomposition of a battery (Swart et al. 2014; Väyrynen and Salminen 2012)

In terms of materials contained in a LIB, Fig. 8.2 shows the average proportions in a pack. They are found in the following:

- Steel and iron: casing of the battery cells, bolts;
- Aluminum and copper: electrode conductors, cables, electronic boards;
- Plastics: casing of the battery, cables, separator;
- Graphite: negative active material of the anode;
- Lithium: electrolyte and positive active material of the cathode;
- Cobalt, Nickel, Manganese, iron, phosphorus: positive active material of the cathode;
- And solvents.

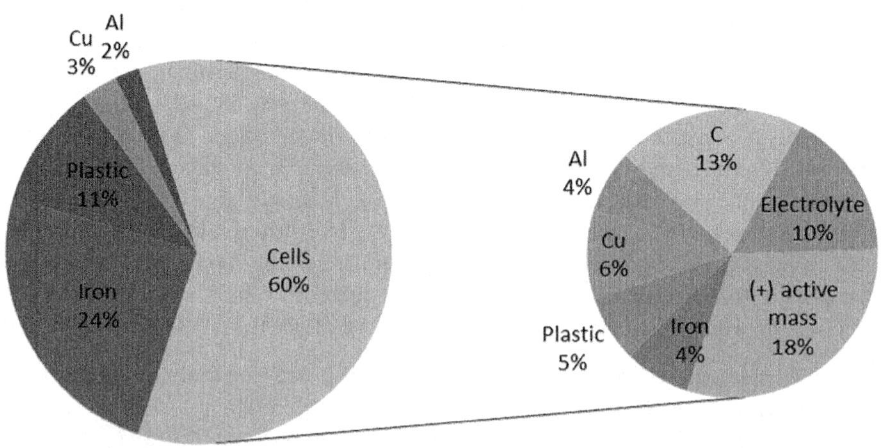

Fig. 8.2 Analytical decomposition of an EV pack

The mass of a LIB is proportional to its capacity, it depends on the vehicle electrification level: hybrid "HEV" (1–2 kWh, about 30 kg), plug-in hybrid "PHEV" (5–15 kWh, about 150 kg) and battery-electric "BEV" (over 15 kWh, about 250 kg).

8.2.2 Recovery Options

Two main LIBs recovery options can be considered: recycling (as required by EU regulation) and repurposing for reuse in 2nd life applications. Due to the evolution of LIB technology and still very low volumes reaching the end-of-life, these recovery options are in their infancy: a structured recovery network does not therefore exist today (since emergent actors, emergent recovery technologies, low volumes, short term contracts between automakers and recyclers).

The next two sections introduce further details on the recovery options plausible.

8.2.2.1 Repurposing for Reuse in 2nd Life Applications

2nd life applications refer to possible uses of repurposed LIBs after the first automotive life. Under the European Li-Ion Battery Advanced Manufacturing (ELIBAMA) project,[1] Törkler (2014) classifies 2nd life applications according to the following three categories:

- Residence related application (3–4 kWh);
- Commercial applications (25 kWh–4 MWh): Telecommunication towers, Light commercial, Uninterruptible power supply (UPS), etc.;
- Energy related/industrial applications (up to 50 MWh): Renewable energy storage, Grid stabilization, etc.

Repurposing for 2nd life use is about dismantling the battery to a defined level, test and replace defective parts, reassemble them into a new configuration and add a battery management system (Ahmadi et al. 2014). Regarding the repurposing processes, they remain to be defined. The main issue is the definition of a good dismantling level: pack, module, cell, or even at the cathode level (Ganter et al. 2014; Georgi-Maschler et al. 2012; Ramoni and Zhang 2013). Precise quantitative data (costs, benefits, advantages, etc.) per process, which as we have shown, is today not even defined. **Thus our repurposing modeling is done generically, regardless of the repurposing process, which will be defined later**.

[1]The objective of the ELIBAMA project is to enhance and accelerate the creation of a strong European automotive battery industry structured around industrial companies already committed to mass production of Li-ion cells and batteries for EV: https://elibama.wordpress.com/.

8.2.2.2 Recycling

The EU directive 2006/66/EC sets the regulatory framework for the treatment of batteries and accumulators at end-of-life. It imposes for EV batteries: (i) The establishment of a collection and a recycling system; and (ii) The requirement to recycle at least 50 % of the battery weight. Therefore the recycling objective is simply the achievement of these regulatory targets, while recovering the value contained in the LIB materials.

These materials are found at different levels in the battery (electrode, cell, pack), in different proportions (remember Figs. 8.1 and 8.2), as well as with a different contribution to the recovered value. Operations of extraction, separation and purification are required consequently. Kwade (2010) identifies four possible basic processing operations: Dismantling; Mechanical conditioning; Pyrometallurgical conditioning; And hydrometallurgical conditioning. These operations could be combined in several ways to form five recycling alternatives. The scientific and industrial state of the art by Idjis (2015) considers two recycling processes as shown in Fig. 8.3. Each process is a succession of three operations, the recovered materials at each operation are described in the bottom. We denote Process 1 by P1 and Process 2 by P2.

We notice that the recovery of materials contained in the positive active material (which are difficult to access) requires an elaborate recycling process, therefore a high cost. This is why materials such as lithium are not recycled today. In the future, with the development of EV, the level of criticality of any material will certainly justify the economic benefit of its recycling, which will reduce this initial criticality.

We have highlighted above the so-called **System Dynamics methodology** (Sterman 2000) i.e. a feedback loop. The concept of feedback loop can be explained using the analogy of vicious or virtuous circles, wherein an influence spreads among several factors and returns back to the factor that initially generated it.

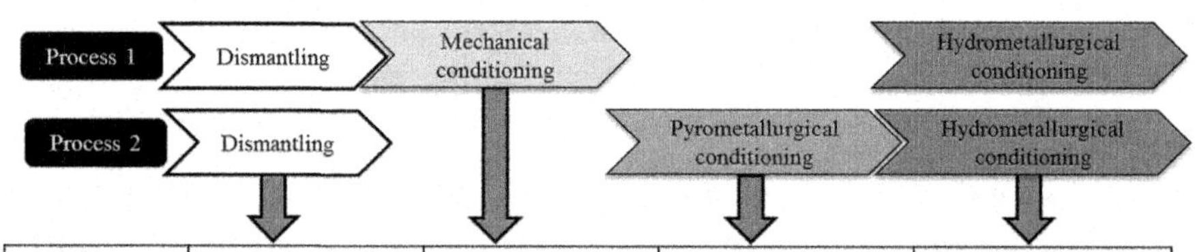

Recuperated elements	metal fractions Plastics	metal fractions Plastics	Alloys Slag	Purified metals (Li, Ni, Mn, Co, Fe)
Treated afterwards	Cells	Active material (+), C, Electrolyte	Alloyss	

Fig. 8.3 Considered recycling processes

8.3 Our Approach

SD is a systems-thinking methodology suitable for the analysis of large-scale complex systems wherein heterogeneous factors interact. “The objective is to analyze, understand and predict the behavior of this system over time by analyzing its changing factors” (Sterman 2000). For our study, this means: (i) Identifying the factors that create the dynamics of end-of-life volumes, recycling profitability, repurposing profitability and end-of-life profitability; (ii) Modeling of internal laws of behavior between these factors and their time simulations in several scenarios. This was done in a much broader study by Idjis (2015), in which other objectives such as material criticality and compliance with recycling targets are investigated (Fig. 8.4).

As shown in the figure above, a SD model is a set of factors related by links of causalities. The feedback loops are represented with the bold arrows and the negative signs indicate that these are negative ones. In the next Fig. 8.5, we will focus on the recycling profitability diagram as one example to explain the whole model.

The diagram is explained as follow: the profitability of any recycling process, ‘P’ is driven by the recycling profit and cost. The recycling cost is calculated based on the volume treated, the initial investment, the fixed cost, the variable cost and the transportation cost; While the recycling benefit is induced also by the treated

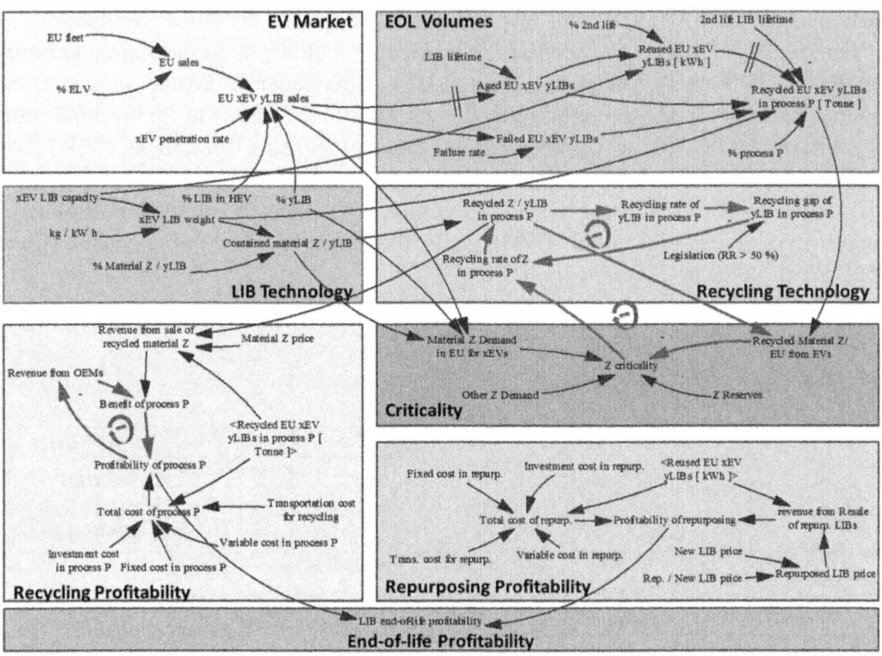

Fig. 8.4 Overview of the LIB recovery network SD model (Idjis 2015)

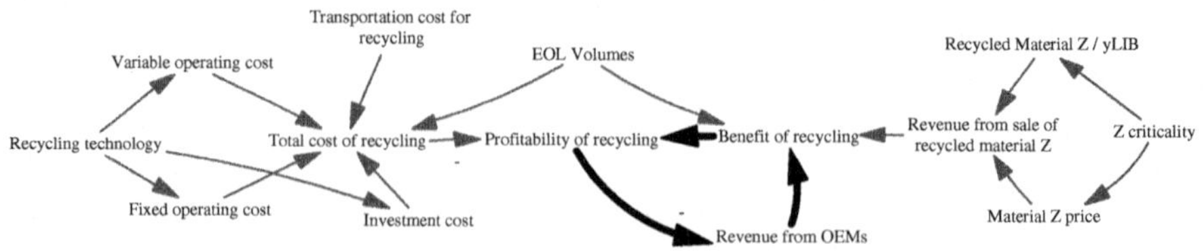

Fig. 8.5 A simplified diagram of the recycling profitability dynamics

volume, the income from the resale of recycled materials and the income from car manufacturers. Finally we have highlighted a feedback loop (in bold, Fig. 8.5) we explain as follow: if the income from the resale of recycled materials is not enough to get the profitability of recycling, the car manufacturer will offset this deficit. Conversely, if the income from the resale of recycled materials is sufficient, the car manufacturer might even receive compensation from this income.

In our prospective study, we have combined the SD with scenarios. The choice of the Stanford Research Institute (SRI) matrix was done because of its suitability with the characteristics of our SD model: complexity, heterogeneous factors and emergence (Acosta and Idjis 2014). The SRI matrix is a crossing of two dimensions of factors that dictate primarily the dynamic evolution of the SD model. The scenarios introduced below are derived from Idjis (2015) (Table 8.1). The first dimension (left side) is about end-of-life volume creation (with two variables) and the second one (top side) is about end-of-life profitability (three variables). Then a particular scenario is a crossing of particular assumptions for these variables.

For example in the S1 scenario, the electric vehicles sales volume is from the IEA's 2DS energy scenario (IEA 2012). This latter is based on proactive environmental policies to contain global warming to 2° (2DS) in 2050, unlike the 4° scenario (4DS). These vehicles have a capacity of 30 kWh and 12.5 kWh for EVs and PHEVs respectively. In the S1 scenario, 80 % of end-of-life automotive LIBs undergo 2nd life reuse before recycling. The main technology is based on nickel, manganese and cobalt (NMC) and the transport cost will decrease from 1500 €/t in 2020 to 1000 €/t in 2030.

Table 8.1 SRI matrix of scenarios

		80%	20%	0%	% 2nd life
		80%	80%→ 50%	80%→ 20%	% NMC
		1500→ 1000 €	1500→ 1250 €	1500 €	Transport cost
12,5 – 30 (kWh)	ETP_2DS	S1	S2	S3	
10 – 24 (kWh)	Mean 2DS - 4DS	S4	S5	S6	
LIB capacity PHEV - EV	EV penetration rate				

As a reminder, we consider a European geographical area. The model is simulated from 2010 (first sales of electric vehicles based on LIBs) until 2050. This is consistent with the literature on EV sales and metals consumption (IEA 2012; Miedema and Moll 2013; Pasaoglu et al. 2012). In this timeframe, the LIBs will be the reference technology to be recovered at least until 2040 and beyond with post LIB batteries.

8.4 Results and Discussion

Section 8.4 analyzes the recycling, repurposing and end-of-life profitability.

8.4.1 Recycling Profitability

In order to analyze recycling profitability, we define the recycling cost and the **global recycling cost that is equal to the recycling cost plus the transport cost for recycling**. These two could be positive or negative: Positive as a source of cost for automakers when the income from the resale of recycled materials does not guarantee the recycling profitability; And negative as a source of profit for car manufacturers.

We calculate a recycling cost that allows a five years return on investment period, and a capital productivity of the order of 2 in 20 years (common industry standards), with a discount rate of 8 % (Hein et al. 2012; Hoyer et al. 2014; Neubauer et al. 2012). This cost will be analyzed in Process 1 (P1) and Process 2 (P2) (see Fig. 8.3) in the six scenarios described above, with sensitivity analysis for each.

Figure 8.6 shows that in 2020, we gat a recycling cost ranking in accordance with the treated end-of-life volumes in each scenario, with a maximum recycling cost of 2500 €/t in S4. For the six scenarios, we observed a decline in the recycling cost between 2020 and 2025 due to economies of scale. From 2025, we notice three trends: Declining for the pair (S1 and S4); Stagnation for (S2 and S5); And increase for (S3 and S6). Cost stagnation and increase in S2/S5, S3/S6 respectively, is due to the increase in the proportion of LFP at the expense of the NMC batteries that have a bigger recycling revenue (NMC = 1569 €/t; LFP = 593 €/t).

As for P1, we show the same trends in P2 in the next Fig. 8.7.

The difference with P1 lies in the initial and the long-term (around 2030) recycling costs. The maximum recycling cost is 5500 €/t for S4 in 2020. The long-term recycling cost stabilizes at 600 €/t in S1/S4; 900 €/t in S2/S5; and 1250 €/t in S3/S6.

The recycling cost in P2 is higher than the one in P1 in between 600 and 700 €/t. This is due to pyrometallurgical conditioning requiring a higher investment (special installations to treat toxic gas from pyrolysis) than the mechanical conditioning.

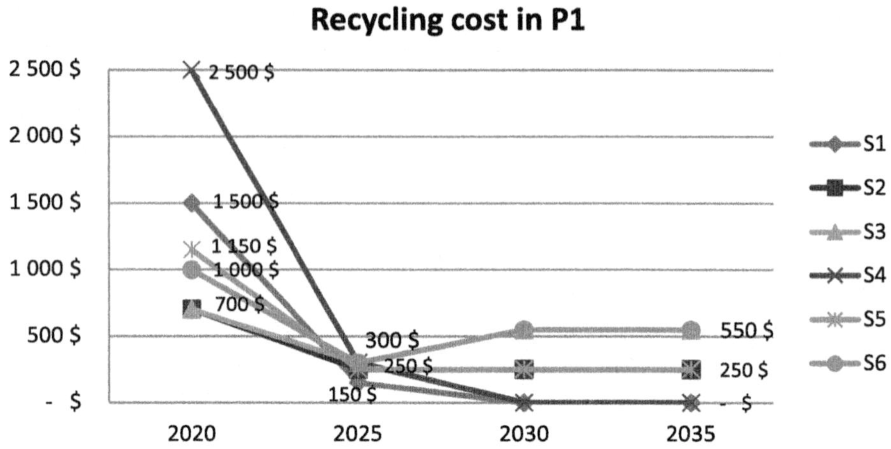

Fig. 8.6 Evolution of the recycling cost in P1 in the six scenarios

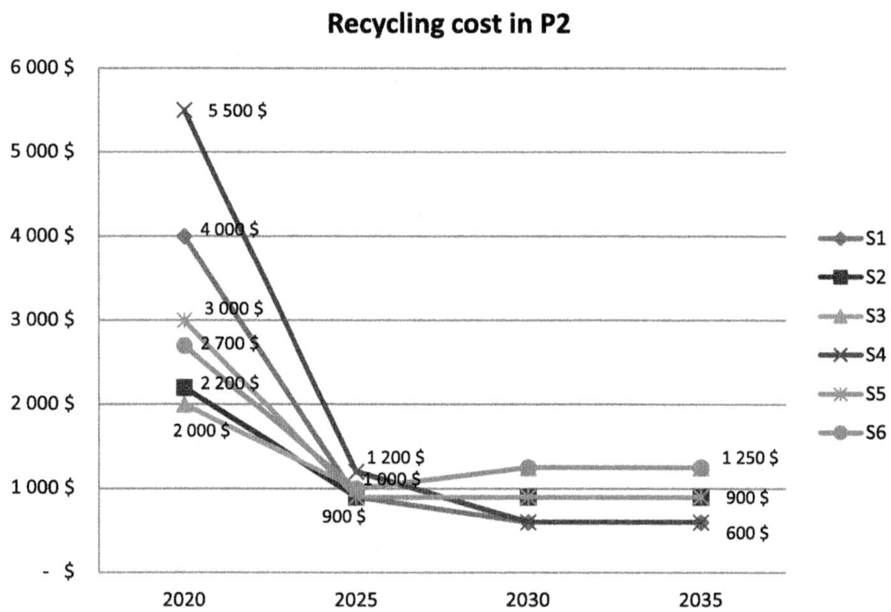

Fig. 8.7 Evolution of the recycling cost in P2 in the six scenarios

As there are other factors (which remained unchanged in the six scenarios) that affect the profitability of recycling, we have proceeded to a sensitivity analysis to investigate the effect of these other uncertain factors. It includes the following:

- Increase and decrease of 50 % in the variable recycling cost;
- Annual increase of 1 and 2 % in the materials resale price;

- Decrease of % 2nd life to 20 and 0 %;
- And the evolution of NMC technology towards nickel or manganese enriched ones.

Considering these four sensitivity analysis, the recycling cost could stabilize in the long term (i.e. 2030) in between −1000 and 1600 €/t, as shown in the following two graphs of Fig. 8.8.

Remind this cost does not include the transportation cost. This latter varies between 1000 and 1500 €/t, depending on the scenario and the period (see Table 8.1). By including it, the overall recycling costs varies in between 0 and 3100 €/t. **We conclude that recycling would remain a net cost in the battery end-of-life, charged to the car manufacturer in all our designed scenarios.**

8.4.2 Repurposing Profitability

The objective is now to assess the conditions for the profitability of this recovery option and its impact on the LIB end-of-life cost. To do so we calculate the repurposing margin as follow: ***repurposing margin (€/kWh) = repurposed LIB selling price – (variable cost of repurposing + transportation cost)***. This latter is of considerable importance because it indicates a potential source of profit for car manufacturers, which could enhance the LIB end-of-life profitability.

Repurposing exists only in scenarios S1/S4 (80 % of end-of-life available volume) and S2/S5 (20 % of volume). For the four scenarios, we have calculated the unprofitability thresholds (Table 8.2) compared to reference values shown in the appendix (Tables 8.8 and 8.9).

When repurposing is profitable, we have a long-term (2030) repurposing margin value between 9.7 and 15.8 €/kWh as shown in the next Fig. 8.9.

We conclude that the repurposing profitability is dependent on the developments of some influencing factors: LIB price, fixed and variable repurposing cost. Under favorable conditions, the repurposing margin is at least 9.7 €/kWh. This could enhance the end-of-life profitability as shown in the next section.

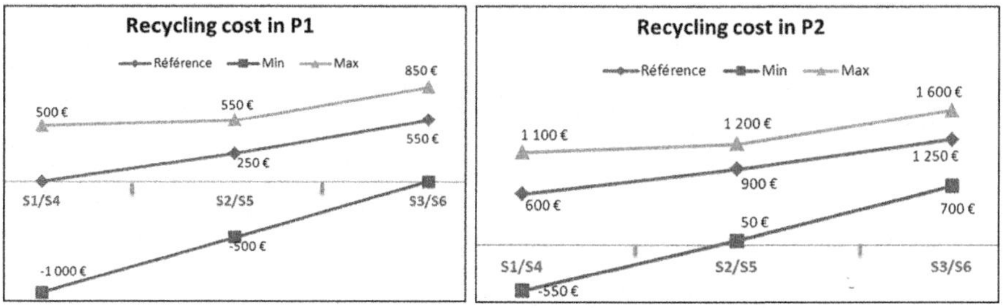

Fig. 8.8 Possible values of the long term recycling cost in both processes

Table 8.2 Repurposing unprofitability thresholds

	S1/S4			S2/S5		
	2020	2025	2030	2020	2025	2030
Repurp. Var cost	> +25 %	>50 %	>50 %	> +25 %	> +25 %	> +25 %
Repurp. Fixed cost	> +50 %	> +100 %	> +100 %	> +25 %	> +50 %	> +50 %
LIB price	< (Lux Research 2015)	< (Lux Research 2015)	</	< (Hein et al. 2012)	< (Hein et al. 2012)	< (Lux Research 2015)

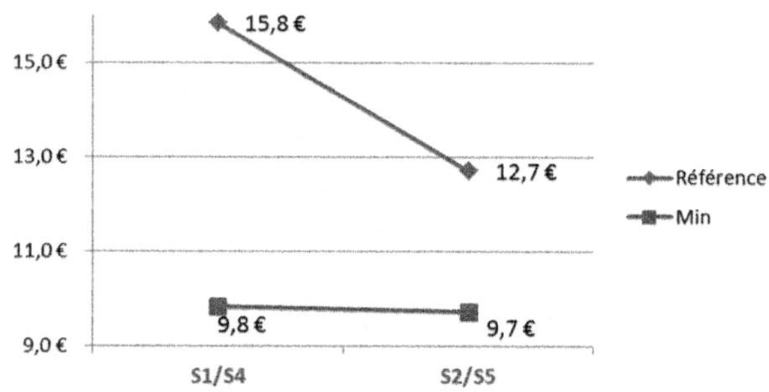

Fig. 8.9 Possible values of the repurposing margin in the long term

8.4.3 *End-of-Life Profitability*

Considering the recycling cost in both processes and the repurposing margin, we calculated the end-of-life cost in the six scenarios. As for recycling, we distinguish the **end-of-life cost** (*recycling cost—repurposing margin*) and the ***global*** **end-of-life cost** (*global recycling cost—repurposing margin*). Looking at the intervals of the recycling cost possible values (Fig. 8.8) and the repurposing margin ones (Fig. 8.9), we obtain the following ranges for the LIB end-of-life cost in the long term (see Fig. 8.10).

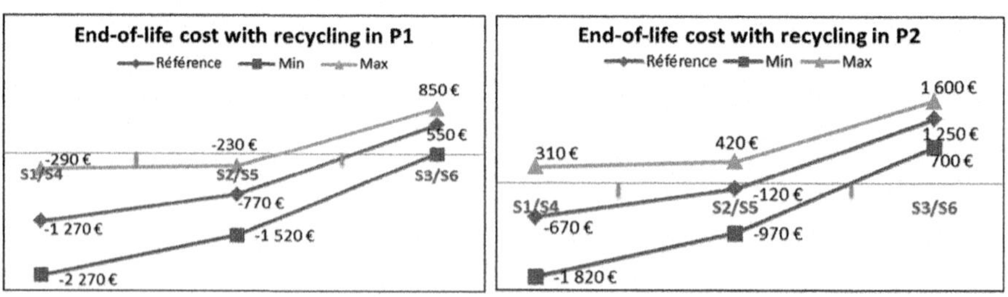

Fig. 8.10 Possible values for the end-of-life cost in the two processes

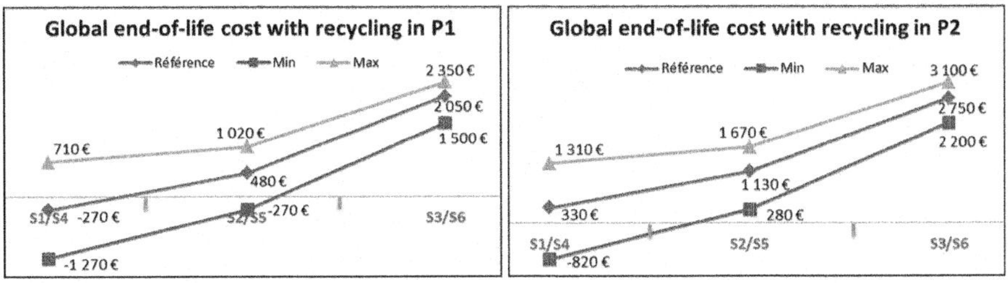

Fig. 8.11 Possible values for the global end-of-life cost in the two processes

Figure 8.10 shows that in P1 (on left), the end-of-life is always beneficial in scenarios S1/S4 and S2/S5, what is not always true in P2. S3/S6 remains a net cost in all situations. If the transportation cost is included into recycling (which is 1000 €/t in S1/S4; 1250 €/t in S2/S5; and 1500 €/t in S3/S6) then we get the following ranges for the **global** end-of-life cost in the long term (see Fig. 8.11).

Comparing P1 and P2 above, we deduce the importance of minimizing the transportation cost because this latter greatly increases the LIB end-of-life cost.

After analyzing the LIB end-of-life cost, we can conclude with the **overall life cycle** cost. This latter includes the LIB initial price and global end-of-life cost (Fig. 8.12). Our interest is to answer the question raised in the introduction: in what extent recovery (recycling/repurposing) is a source of profit or cost in the LIB overall life cycle.

To do this we first analyze the effect of the 'global recycling cost' and the 'repurposing margin' separately, and then together through the 'global end-of-life cost'. Remind of the long-term possible values of these three points in Table 8.3.

In order to harmonize these costs and report them to one LIB, we consider an average LIB with a weight of 250 kg and a capacity of 20 kWh. Regarding the LIB initial price in the long term, we consider an average price of 150 €/kWh (Table 8.9). Therefore we get the following equivalent costs for recycling in Table 8.4. The effect on the LIB overall life cycle cost is calculated in the last column of the table below.

Ultimately, recycling will be neutral in the LIB overall life cycle cost (0 €/kWh) in the best case. In the worst case, recycling will increase the LIB overall life cycle cost by 26 % (39 €/kWh).

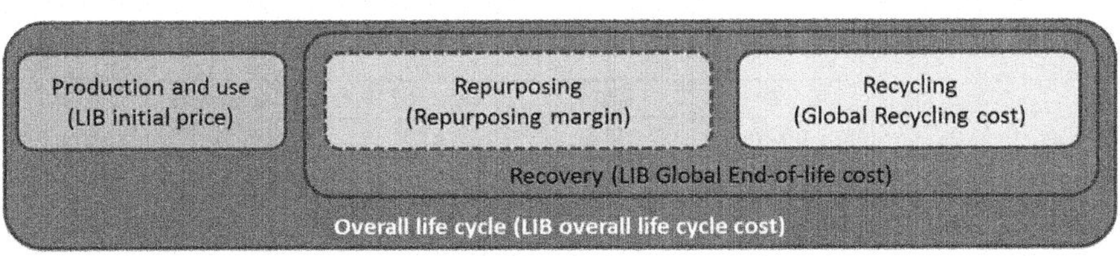

Fig. 8.12 The components of the LIB overall life cycle cost

Table 8.3 Intervals of costs constituting the end-of-life in the long term

	Global recycling cost (€/t)	Repurposing margin (€/kWh)	Global end-of-life cost (€/t)
Min	0	−15.8	−1270
Max	3100	−9.7	3100

Table 8.4 Intervals of the global recycling cost in the long term

	€/t	€/LIB	€/kWh	LIB price increase (%)
Min	0 €	0 €	0 €	0
Max	3100 €	775 €	38.8 €	26

Moving to repurposing, still considering an average LIB of 250 kg and 20 kWh, a future LIB price of 150 €/kWh, we get the following equivalent margins (see Table 8.5). We mention them with a negative sign to mean the reduction effect in the LIB overall life cycle cost (profit).

In the long term, when repurposing is developing, the LIB overall life cycle cost may be reduced by 11 % in the best case. In the worst case, it will be done by just 6 %.

Finally we analyze the effect of the recovery (recycling and repurposing) on the LIB overall life cycle cost, through the LIB global end-of-life cost. Still considering an average LIB of 250 kg and 20 kWh, a future LIB price of 150 €/kWh, we get the following equivalent costs in Table 8.6.

In the long term, and in the best case, where repurposing is developing and the global recycling cost is null, the LIB overall life cycle cost may be reduced by 11 %. In the worst case, corresponding to no repurposing and a global recycling cost of 3100 €/t, the LIB overall life cycle cost may be increased by 26 %.

Table 8.5 Intervals of the repurposing margin in the long term

	€/t	€/LIB	€/kWh	% LIB price increase
Min	**- 1264 €**	**- 316 €**	**- 15.8 €**	**- 11%**
Max	**- 776 €**	**- 194 €**	**- 9.7 €**	**- 6%**

Table 8.6 Intervals of the global end-of-life cost in the long term

	€/t	€/LIB	€/kWh	% LIB price increase
Min	**- 1270 €**	**- 318 €**	**- 15.8 €**	**- 11%**
Max	3100 €	775 €	38.8 €	26%

8.5 Conclusion

In this chapter, we have investigated the recovery of lithium-ion batteries from particularly the end-of-life cost effect on the economy of the LIB and the EV. We have identified two recovery options: (i) Recycling: mandatory by EU regulation, with two main processes (P1 for mechanical and P2 for pyrometallurgical: Fig. 8.3); (ii) Repurposing for 2nd life reuse will only develop if the economic conditions are favorable.

For recycling, we have determined that the cost is higher in the pyrometallurgical process (P2). In 2030, recycling could generate value for car manufacturers in certain conditions (high volume and bigger share of NMC batteries). However, by adding the transportation cost of LIBs for recycling, we only can reach equilibrium (zero cost) in the best case. Hence our results underline the need to properly model and minimize costs of logistics in transportation. The repurposing analysis allowed us to identify the conditions for this recovery option to be profitable. In the long term, repurposing could generate from 9.7 to 15.8 €/kWh of profits. This will significantly enhance the LIB end-of-life profitability.

In the analysis of the end-of-life cost, we have considered simultaneously the two recovery options. We concluded that it is possible that the recovery cannot lower the initial LIB cost, or even increase it by 26 % in the worst case. In the case that it is possible to reduce this initial cost, it is the repurposing that would contribute to it, even if this reduction is weak (11 % of the initial cost in the best case).

Finally we would like to mention a few avenues of research that are worth exploring in the future. As we did with the LIB end-of-life economic cost, future research would consist in assessing the environmental effect of the LIB recovery with life-cycle assessment (ISO 14040:2006). Indeed, the analysis of financial is very relevant to automakers, but the impacts on the environment must be analyzed too, since all stakeholders are facing them.

Acknowledgment This research benefits from the support of the Chair "PSA Peugeot Citroen Automobile: Hybrid technologies and Economy of Electromobility", so-called Armand Peugeot Chair led by CentraleSupélec, ESSEC and sponsored by PEUGEOT CITROEN Automobile.

Appendix

The Appendix highlight the main assumptions used in our SD model, starting from the recycling costs, which are derived from bibliographic research (Hoyer et al. 2013, 2014) and industrial announcements (Umicore 2010) (Table 8.7).

The following Tables 8.8 and 8.9 give the main assumptions for repurposing costs and LIB prices. In this chapter, we have adopted the mean values.

Table 8.7 Recycling costs assumptions

	Dismantling	Mechanical treatment	Hydrometallurgical treatment	Pyrometallurgical treatment
Capacity (t/y)	10,000	6000	4000	7000
Investment	1,850,000 €	3,650,000 €	11,250,000 €	25,000,000 €
Fixed cost €/y	650,000 €	190,000 €	1,500,000 €	1,300,000 €
Variable cost €/t	350 €	250 €	320 €	250 €

Table 8.8 Repurposing costs assumptions

Variable	Cready et al. (2003)	Neubauer et al. (2012)				Mean 4, 10, 20 kWh	SD (%)
		Size (capacity) of repurposed items					
		2 kWh	4 kWh	10 kWh	20 kWh		
Investment $	1,299,778	1,227,097	1,143,457	1,123,737	1,107,777	1,124,990	2
Fixe cost $/y	2,664,315	1,821,921	1,446,063	1,130,599	1,001,900	1,192,854	21
Variable cost $/kWh	22	37	27	21	19	22	21
Capacity kWh/y	73,000	115,000					

Table 8.9 LIB price assumptions

		2015	2020	2025
Avicenne Energy (2014)		390	250	
Hein et al. (2012)		330	225	185
Lux Research (2015)	Tesla	265	195	172
	BYD	295	240	221
	AESC-Nissan	340	260	230
	General motors	360	230	200
Mean		330	233	202
Upper gap		18 %	11 %	14 %
Lower gap		−20 %	−16 %	−15 %

References

Acosta, C., Idjis, H., 2014. State of the art of scenario planning: Proposal of a classification of scenario building existing methods according to use (Mémoire thématique). Ecole Centrale Paris.

ADEME, 2013. Élaboration selon les principes des ACV des bilans énergétiques, des émissions de gaz à effet de serre et des autres impacts environnementaux. Induits par l'ensemble des filières de véhicules électriques et de véhicules thermiques, VP de segment B (citadine polyvalente) et VUL à l'horizon 2012 et 2020. Agence de l'Environnement et de la Maîtrise de l'Energie.

Ahmadi, L., Fowler, M., Young, S.B., Fraser, R.A., Gaffney, B., Walker, S.B., 2014. Energy efficiency of Li-ion battery packs re-used in stationary power applications. Sustainable Energy Technologies and Assessments 8, 9–17.

Alazard-Toux, N., Criqui, P., de Lavergne, J.-G.D., Hache, E., Le Net, E., Lorne, D., Mathy, S., Menanteau, P., Safa, H., Teissier, O., 2014. Les scénarios de transition énergétique de l'ANCRE, Revue de l'Energie.

Avicenne Energy, 2014. xEV Market 2014-2025. Impact on the battery business. Presented at the EEVC, Brussels.

Cready, E., Lippert, J., Pihl, J., Weinstock, I., Symons, P., 2003. Technical and economic feasibility of applying used EV batteries in stationary applications (No. SAND2002-4084). Sandia National Labs., Albuquerque, NM (US); Sandia National Labs., Livermore, CA (US).

EEA, 2015. Global search on data, maps and indicators — European Environment Agency (EEA) [WWW Document]. URL http://www.eea.europa.eu/data-and-maps/find#c1=Graph&c1=Map&b_start=0&c6=transport (accessed 4.14.15).

ERI, 2009. China's Low Carbon Development Path by 2050: Scenario analysis of Energy Demand and Carbon Emissions, Science Press. Energy Research Institue.

European Commission, 2014a. Statistical pocketbook 2014 - Transport. European Commission.

European Commission, 2014b. Regulation (EU) No 333/2014 of the European Parliament and of the Council of 11 March 2014 amending Regulation (EC) No 443/2009 to define the modalities for reaching the 2020 target to reduce CO 2 emissions from new passenger cars.

European Commission, 2011. White Paper on Transport: towards a competitive and resource efficient transport system.

European Commission, 2009. Regulation (EC) No 443/2009 of 23 April 2009 setting emission performance standards for new passenger cars as part of the Community's integrated approach to reduce CO2 emissions from light-duty vehicles.

European Commission, 2006. Directive 2006/66/EC of the European parliament and of the council of 6 September 2006 on batteries and accumulators and waste batteries and accumulators and repealing Directive 91/157/EEC.

Ganter, M.J., Landi, B.J., Babbitt, C.W., Anctil, A., Gaustad, G., 2014. Cathode refunctionalization as a lithium ion battery recycling alternative. Journal of Power Sources 256, 274–280.

Georgi-Maschler, T., Friedrich, B., Weyhe, R., Heegn, H., Rutz, M., 2012. Development of a recycling process for Li-ion batteries. Journal of Power Sources 207, 173–182.

Hein, R., Kleindorfer, P.R., Spinler, S., 2012. Valuation of electric vehicle batteries in vehicle-to-grid and battery-to-grid systems. Technological Forecasting and Social Change 79, 1654–1671.

Hoyer, C., Kieckhäfer, K., Spengler, T.S., 2014. Technology and capacity planning for the recycling of lithium-ion electric vehicle batteries in Germany. Journal of Business Economics.

Hoyer, C., Kieckhäfer, K., Spengler, T.S., 2013. Impact of mandatory rates on the recycling of lithium-ion batteries from electric vehicles in Germany, in: Nee, A.Y.C., Song, B., Ong, S.-K. (Eds.), Re-Engineering Manufacturing for Sustainability. Springer Singapore, pp. 543–548.

Idjis, H., 2015. La filière de valorisation des batteries de véhicules électriques en fin de vie : contribution à la modélisation d'un système organisationnel complexe en émergence (phdthesis). Université Paris-Saclay.

IEA, 2014. Energy Technology Perspectives 2014 - Harnessing Electricity's Potential. International Energy Agency, Paris.

IEA, 2012. Energy Technology Perspectives 2012: Pathways to a Clean Energy System. International Energy Agency, Paris.

IPCC, 2014. Climate Change 2014: Mitigation of Climate Change, Fifth Assessment Report. IPCC Working Group III.

Kwade, A., 2010. On the way to an "intelligent" recycling of traction batteries. Presented at the 7th Braunschweiger Symposium on Hybrid, Electric Vehicles and Energy Management, Braunschweigh.

Lux Research, 2015. Crossing the Line: Li-ion Battery Cost Reduction and Its Effect on Vehicles and Stationary Storage. Lux Research.

McKinsey and Company, 2010. Portfolio of Power-trains for Europe: A Fact-Based Analysis.

Miedema, J.H., Moll, H.C., 2013. Lithium availability in the EU27 for battery-driven vehicles: The impact of recycling and substitution on the confrontation between supply and demand until2050. Resources Policy 38, 204–211.

Neubauer, J.S., Pesaran, A., Williams, B., Ferry, M., Eyer, J., 2012. A techno-economic analysis of PEV battery second use: Repurposed-battery selling price and commercial and industrial end-user value (SAE Technical Paper No. 2012-01–0349). SAE International, Warrendale, PA.

Pasaoglu, G., Honselaar, M., Thiel, C., 2012. Potential vehicle fleet CO2 reductions and cost implications for various vehicle technology deployment scenarios in Europe. Energy Policy 40, 404–421.

Ramoni, M.O., Zhang, H.-C., 2013. End-of-life (EOL) issues and options for electric vehicle batteries. Clean Techn Environ Policy 15, 881–891.

Saveyn, B., Paroussos, L., Ciscar, J.-C., 2012. Economic analysis of a low carbon path to 2050: A case for China, India and Japan. Energy Economics 34, S451–S458.

Sterman, J., 2000. Business Dynamics: Systems Thinking and Modeling for a Complex World. Irwin/McGraw-Hill, New York.

Swart, P., Dewulf, J., Biernaux, A., 2014. Resource demand for the production of different cathode materials for lithium ion batteries. Journal of Cleaner Production 84, 391–399.

Tian, W., Da Costa, P., 2014. A Sectoral Prospective Analysis of CO2 Emissions in China, USA and France, 2010-2050. Presented at the The 4th IAEE Asian Conference. IAEE, Beijing, China.

Törkler, A., 2014. Electrodes and cells manufacturing white paper:Batteries Refurbishing & Reuse. ELIBAMA.

Umicore, 2010. The UMICORE Process: Recycling of Li-ion and NiMH batteries via a unique industrial Closed Loop.

Väyrynen, A., Salminen, J., 2012. Lithium ion battery production. The Journal of Chemical Thermodynamics, Thermodynamics of Sustainable Processes 46, 80–85.

Chapter 9
Transmission System Operator Regulation for Electric Vehicle Fleets: A Survey of the Issues

Yannick Perez and Marc Petit

Abstract A modular framework is used to analyze how Grid Integrated Vehicles (GIVs), i.e. bi-directional plug-in electric vehicles that are able to modulate their charging rate and have bi-directional capabilities, could be managed efficiently to deliver grid services for transmission operators and conversely, how these new services could be set aside by the design of the current rules in some regions. Based on a detailed analysis of the rules implemented by some representative TSOs, we discern two modules that gather the essential rules for GIV development: the *rules towards aggregation of EVs*, and *the rules defining the payment scheme* of the services provided by GIVs. We deduce an optimal combination among these rules that could define the ideal organization for GIVs. Finally, we confront this ideal TSO organization with the European guidelines under construction.

Keywords Electric vehicles · Transmission system operator · Market design

9.1 Introduction

It is puzzling today to analyze the organizational diversity of Transmission System Operators (TSOs) in competitive electricity markets around the world (Rious et al. 2008). In liberalized power systems, TSOs present different organizational forms or governance structures, depending on network ownership, on applied regulation, etc. In order to understand this wide range of organizational forms, we argue that the

Y. Perez (✉)
LGI Centrale Supélec Paris, Helmholtz Centre for Environmental Research—UFZ, RITM Université Paris-Sud, Permoserstr. 15, 04318 Leipzig, Germany
e-mail: yannick.perez@u-psud.fr

M. Petit
GeePs, CNRS UMR 8507 CentraleSupélec, UPSud and UPMC, 91190 Gif-sur-Yvette, France
e-mail: marc.petit@supelec.fr

D. Attias (ed.), *The Automobile Revolution*, DOI 10.1007/978-3-319-45838-0_9

use of Wilson (2002) and Baldwin and Clark (2002) modular frame is helpful in defining and classifying the rules of the game from the most to the least adapted to new services and innovations.

Considering research dealing with Grid Integrated Vehicles (GIVs), i.e. plugin electric vehicles that are able to modulate their charging rate and have bidirectional capabilities, coalitions of electric vehicles (EVs) are likely to be integrated into TSO reserves in the future (Han et al. 2010; Kempton and Letendre 1997; Kempton and Tomić 2005a). In this respect, GIVs will become Reserve Providing Units, which are required and financed by the TSO set of rules that are today very differently organized around the world.

In this paper, we apply the aforementioned modular frame to analyze how GIVs could be managed efficiently to deliver valuable services for transmission operators and conversely, how these new services could be set aside by the design of the current rules in some regions.

This work builds on (Codani et al. 2014) which already identified the representative TSO rules for GIVs providing frequency control. In this paper, the authors make a comparison among five representative TSOs on a list of rules and characteristics that are important for GIV deployment. The five TSOs in question, and the regulation manuals associated are: Energinet.dk (2012), RTE (2004, 2011a, b), ERCOT (2012, 2013a, b, c, d), CAISO (2010, 2011, 2013) and PJM (2012, 2013a, b, c). The National Grid Company of the UK has been added in this paper to increase the diversity of possible rules regarding demand side response (DSR) regulation and remuneration (National Grid 2012a, b, 2013). The term DSR encompasses bidirectional demand services, which will enable to take full advantage of EV potential.

Based on this previous literature, we go a step further and identify two key modules that gather the essential rules for GIV development: the rules towards aggregation of EVs, and the rules defining the payment scheme of the services provided by GIVs. The novelty in this work relies on a deeper and more policy-oriented analysis of the most important rules detailed in Codani et al. (2014). Moreover, we analyze the proposed ENTSOE network codes, in order to shed some light on the possibility of implementing the ideal TSO in reality.

The paper is organized as follows. Sections 9.2 and 9.3 are dedicated respectively to the identification of the best aggregation rules and payment schemes for grid services. In Sect. 9.4, we propose an ideal TSO organization, and compare it to the 6 representative TSOs understudy. In Sect. 9.5 we discuss some policy recommendations by screening European issues and ENTSOE network codes through the lenses of our framework. Last section concludes our main findings.

9.2 Module 1: Aggregation Rules

An aggregator has a fundamental role in GIV architectures for TSO services: it is responsible for presenting a fleet of EVs as a single entity to the TSOs. Aggregators are required because: (a) TSOs deal with large entities (MW rather than kW size),

(b) TSO data processing capabilities do not have the bandwidth for controlling millions of kW size units; they were designed for 100s of multi-MW sized units, and (c) TSOs expect their resources to be reliable, which is a problem for a single EV. An EV necessarily gives first priority to transportation, but from the power system perspective one EV may leave the power system at any moment. Aggregators can address these issues by controlling a large number of EVs (Kamboj et al. 2011; Kempton and Tomić 2005b), and offering a single, statistically-reliable entity to the TSO. Finally, aggregators should also be able to deal with a large diversity of degrees of information and degrees of uncertainty induced by many different vehicle types, driver plans, and regularities in driver behaviors (Kempton and Letendre 1997; Bessa and Matos 2010), details well outside the business expertise or interest of TSOs.

Correspondingly, TSOs must allow such aggregation for GIV use, but what are the main rules to do so? Here we would like to insist on three rules: the size of the minimum bid, the interoperability among DSOs, and the technical form of aggregation.

9.2.1 The Size of the Bid

In all reserve markets, bids cannot be less than a minimum power level; we have seen a range of minima from 100 kW (PJM, frequency regulation) to 10 MW. In terms of EV coalitions, this minimum-bidding amount can be converted into a minimum number of EVs. A high level of minimum bidding amount would represent a challenge for the development of pilot and early commercial projects, because they may not have enough vehicles to meet with the minimum.

For instance, considering charging stations of 3 kW (domestic plugs), and that one EV out of three is available for reserve markets (because of transportation, charging needs), the minimum fleet size would be 100 vehicles for a minimum bid value of 100 kW. On the other side, given a minimum bidding size of 10 MW, the number of EVs in a coalition should be at least 10,000. These figures should be put in perspective with those of today EV sales; in 2013, only around 10,000 EVs were sold in France for example. Thus, given a high minimum bidding size, it would be impossible to make a coalition of privately owned vehicles in France, not to mention a company fleet.

Even if we consider a high penetration of EVs, say, in 10 years, a high minimum bidding value would narrow the diversity of potential aggregators: among others, company fleets of utility vehicles, or Vehicle-to-Building scenarios (fleets gathering vehicles parked in the same parking lot), would not be allowed to become aggregators.

9.2.2 The Scope of the Bids

Single or multiple DSO zones of technical regulation are a second key concern for aggregation business. As EVs are small moving storage entities, TSOs rules should also allow resources that may shift locations, and may be spread across electrical distribution companies (EDCs). One way to manage that may be to register charging stations rather than registering EVs. Regarding movement across EDCs, some TSOs work with very few EDCs (for instance RTE, whose main EDC partner is ERDF) but others are connected to many of them (for instance Energinet.dk, with 65 EDCs), and in the latter, more typical case, not being allowed to aggregate across EDCs makes aggregation more challenging or even impossible.

Thus for our modular frame, the best option is to allow and organize interoperability among various DSOs as done in RTE or Energinet.dk. From an aggregator point of view, a restrictive implementation of this rule could be very constraining. Indeed, the minimum fleet size is induced by the rationale described in Sect. 9.2.1, and if this minimum has to be reached in a single DSO area, it may be impossible for aggregators to meet with the minimum fleet size requirements.

9.2.3 The Precision of the Action

Our last criterion is a distinction between telemetry and financial aggregations. Telemetry is the desired form of aggregation; it enables combining bids and then controlling distributed power flows from one or more central locations. In contrast, financial aggregation only allows combining financial bids but not power flows, which, among other things, would prevent an aggregator from implementing dispatching algorithms.

In our frame of combining the best rule issued from the TSOs survey, we propose to select the PJM and Energinet.dk rules that allow the telemetry aggregation. It is noticeable that TSOs may be unwilling to allow large telemetry aggregations, because it would make verification of reserve activation more difficult.

9.2.4 Partial Conclusion

The Table 9.1 sums up the identified rules regarding aggregation and the different possible organizations for each rule.

Table 9.1 The different organizations for Module 1

Aggregation rule module	Organization	
	Best option	Restrictive option
Size of the bid	100 kW	10 MW
Interoperability among DSOs	Possible	Impossible
Aggregation level	Telemetry	Financial

9.3 Module 2: The Rules Defining the Payment Scheme of Grid Services

9.3.1 Regulated of Market Based?

For a given reserve market, TSOs may differ in their way of dispatching the required power among all the units that are part of the reserve in question (whether primary or secondary). Raineri et al. (2006) identified several transaction mechanisms. Some TSOs implement open markets in which units are allowed to bid as they want in these markets, a bid being an amount of offered capacity and its associated price. Bids are either accepted or rejected by the TSO. Other TSOs dispatch the total required power among all the units that are part of the reserve in proportion to their historical load share. In this situation, depending on the TSOs, providing reserve is either a choice or compulsory for a unit.

As a partial conclusion in our framework, the use of market prices as a way to determine the dispatch of reserves is much more profitable for new storage resources such as GIVs. First, in a regulated approach, we would have to wait for a formal change of the rules, so that they would be suitable for new resources such as GIVs. This adaptation is likely to be lengthy, sub-optimal, and lagging EV sales, trying to catch up with the market evolution instead of taking the lead. Then, changing a contract binding an aggregator and the TSO would also be a lengthy process, one that is not really compatible with a dynamic EV fleet. Indeed, in addition to many dynamic changes within a fleet because of transportation needs, the EV fleet itself is also likely to evolve, with new EVs joining or leaving the program.

9.3.2 Complete or Incomplete?

Besides the nature of the payment scheme, the second element of our frame is its consistency regarding the services offered or possibly offered by EVs. It is puzzling to identify ancillary services that are required but not remunerated specifically by some TSOs and DSOs. Some services are just mandatory with no explicit remuneration nor explicit reserve allocation method. Examples are for instance PJM or CAISO not paying for primary frequency regulation.

Regarding our analysis, the more the payment scheme is incomplete and does not compensate for the services provided by all the actors, the more EVs are penalized in their contribution as GIV resources. A clear and complete payment scheme is needed as a condition in the Ideal TSO we seek to build.

From the TSO perspective, it could be beneficial to complete the payment scheme of ancillary services. Indeed, as AS providing units do not have any incentive to provide these unpaid services, they sometimes achieve poor performance in the provision of these services. For instance, Ingleson et al. (2009) point out the fact that the total frequency droop (also referred to as frequency characteristic) of Eastern Interconnection in the US has been dangerously decreasing for the past 10 years, jeopardizing grid security.

9.3.3 Additional Bonus for Intense Flexibility Providers?

In the United States, the Federal Energy Regulatory Commission (FERC), a federal agency responsible for harmonizing interstate energy laws and TSO rules, has investigated the different frequency regulation compensation practices of TSOs in (Federal Energy Regulatory Commission 2011). Its conclusion is that current compensation methods are unjust and discriminatory, specifically because fast ramping resources (resources that are able to change their output very quickly) are not remunerated enough with respect to the greater amount of frequency regulation provided.

To deal with this issue, the FERC makes two recommendations. First, remuneration should not only be based on availability (i.e. in \$/MW), but also on utilization (\$/MWh), and every MWh exchanged with the grid for the purpose of frequency control should be counted as a source of positive revenue for the resource in question, whether the MWh flowed from the grid to the resource or from the resource to the grid. That way, as fast-ramping resources respond faster, they exchange more MWh with the grid than slow-ramping units, so payment will be fairer.[1]

Second, regulation resources should receive a two-part payment: the first one is the capacity and utilization payment discussed above, including an opportunity cost, and the second one is based on performance, taking into account the response accuracy. Further details about the performance calculation are provided in the more recent FERC order 784 (Federal Energy Regulatory Commission 2013a) speed and accuracy should be taken into account in the payment of ancillary services.

[1]If such a dual payment is implemented, the net metering of energy flows, i.e. the fact that energy flowing from the EV to the grid would be paid the same price as the energy flowing from the grid to the EV, is not required any more.

EVs are very fast-ramping resources. Therefore, TSOs that abide by FERC compensation recommendations, or similar compensation schemes reflecting the value of fast responses, are more attractive for GIV aggregators. Considering our optimal TSO, the best solution is then to be able to benefit from this kind of bonus. However, the implementation of this bonus should be managed carefully. Indeed, the addition of an extra bonus to an existing payment scheme should be set at the efficient level. The risk induced by introducing a bonus is that it might create a distortion that could either overcompensate the initial problem, or not compensate enough and leave the issue unsolved.

An alternative way of proceeding would be to regard fast and slow ramping tenders as two different products. Thus, establishing a separate market earmarked for fast-ramping resources, with its own rules and regulations, might be another solution to remunerate these services in a just and fair manner. At last, some electrical grids might not presently feel the need for fast-ramping resources. The droop control method of conventional units has been operating for a long time and seems to be working quite well (provided that financial incentives are adequate, see Sect. 9.3.2). Thus, introducing faster responses in this context has to be investigated thoroughly by the TSOs. In a first time, such services may be mostly suited to extreme frequency containment plans after severe disturbances rather than for normal operations. Then, with the increasing penetration of intermittent renewable sources, which induce more production fluctuation and less system inertia, fast-ramping units may be more and more required. We are already observing this phenomenon in some island networks, which are isolated—so very sensitive to frequency drops—but benefit from substantial wind and solar resources.

9.3.4 Partial Conclusion

The Table 9.2 sums up the rules dealing with the payment scheme, and the different possible organizations.

Table 9.2 The different organizations for Module 2

Payment scheme module	Organization	
	Best option	Restrictive option
Nature of the payment	Market based	Regulated
Form of the payment	Complete scheme	Incomplete scheme
Bonus for flexibility	Set at the efficient level, or separate market created	Not existing

9.4 The Ideal TSO for Grid Integrated Vehicles

The optimal implementation of the TSO rules for GIVs providing Demand Side Response, resulting from a combination of the organizations previously presented, is displayed in Table 9.3. Within this ideal TSO, there is almost no barrier to the building of EV coalitions. The latters can be part of all reserves, which are all remunerated in a fair way. Therefore in this TSO, small company fleets are aggregated and participate into reserve markets. Similarly, privately owned parking open to public (in malls, or commercial buildings) can aggregate their charging stations and offer their customers to provide grid services. These two kinds of fleets could provide local DSO services, such as load shifting or voltage regulation, as well as TSO services. Then, with the increasing EV penetration, larger fleets, participating to widespread TSO services spread over multiple DSO areas, would be aggregated. They could take advantage of being very statistically reliable, though still very efficient.

It is noticeable that this table results from an analysis of representative TSOs. We cannot logically exclude the fact that some non-studied TSOs may enjoy even more favorable rules that would be missing in our ideal TSO. However, we chose a wide diversity of TSOs, including some at the forehead in terms of regulation for new technologies and smart grids. Thus, despite the fact that we may have missed some particular rules, we would like to show that using this comparative frame already deliver some valuable insights.

In order to do so, we compare the TSOs understudy with the ideal one. Results are presented in Table 9.4.

Table 9.3 Ideal TSO organization

Rule	Best organization
R1: Size of the bid	100 kW
R2: Scope (multiple DSO)	Possible
R3: Aggregation level	Telemetry
R4: Nature of the payment	Market based
R5: Consistency of the payment	All AS services should be paid
R6: Bonus for extra flexibility	Set at the efficient level, or separate market created

Table 9.4 Evaluation of the representative TcSOs

TSO	R1	R2	R3	R4	R5	R6
RTE	X	√	√	X	√	X
PJM	√	X	√	√	X	√
ERCOT	√	X	X	√	X	X
Energinet.dk	~	√	√	√	√	X
CAISO	~	X	X	√	X	√
NGC	X	√	√	~	X	X

√: good ~: so so X: bad

We can infer two main conclusions from this table. First, there is no TSO implementing a perfect regulation favorable to the development of GIV. However, some of them are closer to the ideal one than others. Second, our frame can be used as a methodological tool, which can be applied to other TSOs in order to assess their friendliness towards GIV deployment, and guide reforms towards what should be done to go a step further.

9.5 Policy Recommendations Towards ENTSOE

In this part, we try to see whether the ideal TSO as described in Sect. 9.4 is reachable with ENTSOE recommendation, which are provided in the Network Codes. The latter's are still under revision, but recent versions are publicly available. They should come into force in the following years, guiding TSOs in their organizations.

The new ENTSOE network codes introduce the concept of Significant Grid User (SGU) (ENTSOE 2013a). The latter could be either a demand facility connected to the DSO, or an aggregator as defined in ENTSOE (2012), that is, a legal entity responsible for the operation of a number of demand facilities by means of demand aggregation. SGUs should be allowed to provide Demand Side Response (DSR) directly to the TSO, or via the DSO. (ENTSOE 2012) also suggests the implementation of DSR Very Fast Active Power Control (DSR VFAPC), defined as a very fast (within 2 s) demand modulation in response to frequency fluctuations.

Moreover, TSOs should avoid undue barriers for new entrants, and make the participation to DSR for storage units easier (ENTSOE 2013b). With respect to the ENTSOE role model (ENTSOE 2011), it means that aggregators and energy storage units should be allowed to become Balance Responsible Parties (BRPs).

According to ENTSOE guidelines, units providing Frequency Containment Reserves (FCR) (also referred to as frequency response or primary control) should have the right to perform telemetry aggregation if the coalition size does not exceed 1.5 MW (ENTSOE 2013c).

The ENTSOE network code on Electricity Balancing (2013b) specifies that the procurement of Balancing Services should be market based and market based only.

Some rules are not addressed within the network codes because they are left at the discretion of each TSO. We can see that there is a good correlation between the ENTSOE guidelines and the ideal TSO; the network codes are paving the way for the implementation of a complete DSR framework suitable for all new controllable loads.

Table 9.5 makes a comparison between our ideal TSO and the ENTSOE future rules and guidelines.

Table 9.5 Ideal TSO versus ENTSOE guidelines

Rule	Ideal TSO	ENTSOE proposals
Minimum size	100 kW	Not addressed
Interoperability among DSOs	Possible	Not clearly defined, but TSOs and DSOs should make all endeavors and cooperate in order to ease the participation to DSR
Aggregation level	Telemetry	Status of *aggregator* defined. Telemetry aggregation considered for FCR up to 1.5 MW
Nature of the payment	Market Based	Market based
Incompleteness of the payment	All AS should be paid	All AS should be paid
Extra bonus for flexibility	Set at the efficient level/separate market created	DSR VFAPC should be implemented

9.6 Conclusion and Future Work

In this paper, we used a modular frame, composed of two different modules, to define an ideal TSO for GIVs providing Demand Side Response. The rules of this ideal TSO would encourage the formation of EV coalitions, no matter neither their sizes nor their geographical expansions. All TSO services would be remunerated in a fair and just manner. The comparison between some representative TSOs and the ideal one reveal that some of them are not far from implementing the optimal combination of rules. The guidelines provided by ENTSOE network codes should strengthen this trend as the analysis of these documents outline their correlations with the ideal TSO.

Our modular frame is of course not exhaustive. For instance, we did not deal with the status of energy storage units, which is not properly defined by each TSO (Codani et al. 2014) [although the FERC is pushing forward for the integration of these new units (Federal Energy Regulatory Commission 2013b)]. Further work could then consist in completing this study framework.

Acknowledgments The authors gratefully acknowledge the contributions of Bjoern Christensen (Nuvve), Ole Jan Olesen and Yannick Phulpin (EDF R&D).

This research benefits from the support of the Chair "PSA Peugeot Citroen Automobile: Hybrid technologies and Economy of Electromobility", so-called Armand Peugeot Chair led by Ecole Centrale Paris, ESSEC and SUPELEC and sponsored by PEUGEOT CITROEN Automobile.

References

C. Y. Baldwin and K. B. Clark, 2002. "The Option Value of Modularity in Design: An Example from Design Rules, Volume 1: The Power of Modularity," *SSRN Electronic Journal*, [Online]. Available: http://www.ssrn.com/abstract=312404.

R. J. Bessa and M. A. Matos, 2010. "The role of an Aggregator Agent for EV in the Electricity Market," in 7th Mediterranean Conference and Exhibition on Power Generation, Transmission, Distribution and Energy Conversion, no. November. IET, p. 9.

CAISO, 2013. "Tariff Clarifications to Implement Pay for Performance Regulation".

CAISO, 2011. "Non-Generator Resources - Regulation Energy Management - Frequently Asked Questions".

CAISO, 2010. "Proposal for participation of non-generator resources in California ISO ancillary services markets".

P. Codani, K. Willett, and L. Andrew, 2014. "Critical rules of transmission system operators for grid integrated vehicles providing frequency control," working paper.

Energinet.dk, 2012. *Ancillary services to be delivered in Denmark Tender conditions.*

ENTSOE, 2011. The Harmonised Electricity Market Role Model.

ENTSOE, 2012. Network Code on Demand Connection, [Online]. Available: https://www.entsoe.eu/major-projects/network-code-development/demand-connection/

ENTSOE, 2013a. Network Code on Operational Security, [Online]. Available: https://www.entsoe.eu/major-projects/network-code-development/operational-security/

ENTSOE, 2013b. Network Code on Electricity Balancing, [Online]. Available: https://www.entsoe.eu/major-projects/network-code-development/electricity-balancing/

ENTSOE, 2013c. Network Code on Load-Frequency Control and Reserves, [Online]. Available: https://www.entsoe.eu/major-projects/network-code-development/load-frequency-control-reserves/

ERCOT, 2012. Nodal Protocols Section 8 : Performance Monitoring.

ERCOT, 2013a. Nodal Protocols Section 6 : Adjustment Period and Real-Time Operations.

ERCOT, 2013b. Nodal Protocols Section 4 : Day-Ahead Operations.

ERCOT, 2013c. Nodal Protocols Section 3 : Management Activities for the ERCOT System.

ERCOT, 2013d. Nodal Protocols Section 16 : Registration and Qualification of Market Participants.

Federal Energy Regulatory Commission, 2011. "Order 755 - Frequency Regulation Compensation in the Organized Wholesale Power Markets".

Federal Energy Regulatory Commission, 2013a. "Order 784 - Third Party Provision of Ancillary Services; Accounting and Financial Reporting for New Electric Storage Techno".

Federal Energy Regulatory Commission, 2013b. "Order 792 - small generator interconnection agreements and procedures".

S. Han, S. Member, S. Han, and K. Sezaki, 2010. "Development of an Optimal Vehicle-to-Grid Aggregator for Frequency Regulation," *IEEE Transactions on Smart Grid*, vol. 1, no. 1, pp. 65–72.

J. W. Ingleson, S. Member, E. Allen, D. E. Interconnection, and A. Beta, 2009, "Tracking the Eastern Interconnection Frequency Governing Characteristic," in 2010 IEEE Power and Energy Society General Meeting, pp. 1–6.

S. Kamboj, W. Kempton, and K. S. Decker, 2011. "Deploying Power Grid-Integrated Electric Vehicles as a Multi-Agent System," 10th Int. Conf. on Autonomous Agents and Multiagent Systems - Innovative Applications Track (AAMAS), vol. 1, no. Aamas, pp. 2–6.

W. Kempton and S. Letendre, 1997. "Electric Vehicles as a new power electric utilities source for electric utilities," *Transportation Research Part D: Transport and Environment*, vol. 2, no. 3, pp. 157–175.

W. Kempton and J. Tomić, 2005a. "Vehicle-to-grid power fundamentals: Calculating capacity and net revenue," *Journal of Power Sources*, vol. 144, no. 1, pp. 268–279.

W. Kempton and J. Tomić, 2005b. "Vehicle-to-grid power implementation: From stabilizing the grid to supporting large-scale renewable energy," Journal of Power Sources, vol. 144, no. 1, pp. 280–294.

National Grid, 2012a. Balancing code 3: Frequency control process.

National Grid, 2012b. CUSC - SECTION 4 BALANCING SERVICES.

National Grid, 2013. Grid Code Chapter 06 - Connection conditions.

PJM, 2012. Manual 12: Balancing Operations.

PJM, 2013a. Manual 1: Control Center and Data Exchange Requirements Control Center Requirements.

PJM, 2013b. Manual 10: Pre-Scheduling Operations,

PJM, 2013c. Manual 11 : Energy & Ancillary Services

R. Raineri, S. Rios, and D. Schiele, 2006. "Technical and economic aspects of ancillary services markets in the electric power industry : an international comparison," Energy Policy, vol. 34, no. 1010750, pp. 1540–1555.

V. Rious, J.-M. Glachant, Y. Perez, and P. Dessante, 2008. "The diversity of design of TSOs," *Energy Policy*, vol. 36, no. 9, pp. 3323–3332.

RTE, 2004. "Mémento de la sûreté du système électrique".

RTE, 2011a. Documentation Technique de Référence Chapitre 4.1 – Contribution des utilisateurs aux performances du RPT.

RTE, 2011b. Documentation technique de référence Article 8 . 10 – Modèle de contrat de participation aux services système.

R. Wilson, 2002. "Architecture of Power Markets," *Econometrica*, vol. 70, no. 4, pp. 1299–1340, [Online]. Available: http://doi.wiley.com/10.1111/1468-0262.00334.

Chapter 10
Conclusion

Danielle Attias

At the Universal Exhibition in 1889, the first steam car was presented by Serpollet Peugeot. This amazing invention has had considerable resonance throughout society. It resulted in a total upheaval of individuals over space and time, a new era of freedom, of pleasure and a luxury symbol. The automotive industry was soon after born as one of the most successful economic and industrial sectors of the twentieth century. Its growth showed us that the transport of people and goods can be thought differently. And the car makes us dream.

120 years later, a research team of Stanford's University develops a driverless vehicle, thus sending the second revolution of the automobile underway. The car becomes intelligent and 'transports' us effortlessly, freeing our time, offering us more safety and a new comfortable lifestyles. Hybrid, electric, automatic and autonomous vehicles are all scenarios that are emerging before us and the challenges for the global automotive industry are considerable. Rethinking the future car is to think about the future of mobility, imagine tomorrow's city and the lifestyle of its millions of inhabitants.

Will the automotive industry be sufficiently agile and ingenious to meet these challenges? The arrival of newcomers from other industries, desperate to gain market shares and new customers causes a disruptive change in the automotive value chain. An industrial and cultural revolution is now beginning for all stakeholders.

To analyze this automotive revolution, we have chosen to be situated within the framework of a global approach of mobility. The study of multi-factors, economic, social, industrial, energetic and political have helped to better understand this paradigm shift. The very concept of mobility is changing in nature. New needs, new

D. Attias (✉)
Industrial Engineering Research Department/Laboratoire Génie Industriel,
CentraleSupélec, Université Paris-Saclay, 92290 Chatenay-Malabry, France
e-mail: danielle.attias@ecp.fr

D. Attias (ed.), *The Automobile Revolution*, DOI 10.1007/978-3-319-45838-0_10

attitudes and usages lead to the emergence of a new mobility model and to a deep transformation of society as a whole.

Indeed, with the advent of new vehicles and new services, the buildings and cities infrastructure recomposes and joins a new ecosystem based on the increasing importance of information and communication technology. This technological innovation is necessarily correlated to the social and economic model in which it operates. Being connected permanently to the world became, for the younger generations, more important than the possession of a vehicle. In the digital economy, user expectations are more targeted, it is both optimizing transportation time and reducing travel costs.

A new reflection is committed to the urban space and the national or local policies that accompany it. The study of *smart cities* reflects this change because these new spaces have in common the data management optimization in order to improve urban services: transport, energy, waste, housing, health, education and culture. In response to congestion in cities and endemic pollution, it will not be possible to adapt transport systems without changing the cities themselves. Rearranging human activities, rationalizing the flows of people and goods is an ongoing challenge. It is in this perspective that the debate lies in the transformation of cities into Smart Cities. However, the various transport options available to us, as intermodality or car-sharing, are complex to the extent that they strongly impact the structuring of public spaces. It should also be noted that mobility is not yet fair for all. Everywhere in the world, the gap remains between megacities with diverse and abundant mobility and suburban or rural areas where multi-modality opportunities are more limited and sometimes non-existent.

The period we are going through is, without a doubt, a transitional period. In the 2020/2025 horizon, the urban landscape will have completely changed. Traditional cars will coexist with environmentally friendly vehicles, maybe even autonomous ones and new questions will arise as to the management of all these types of vehicles flows and regulation. Another challenge is the geostrategic one. It will involve the supply of rare earth and strategic materials, such as lithium, to the extent that we are increasingly dependent on producer countries and that we need to meet demand in the coming years.

Finally, what will be the level of acceptability of the user/customer of this extraordinary automobile revolution? Ultimately, customers will undoubtedly be those who will determine the technological choices of the automotive industry and guide the direction of history.

Printed by Printforce, the Netherlands